Saving our Planet

Jihad Saffouri

Published by Jihad Saffouri, 2025.

SAVING OUR PLANET

First edition. April 14, 2025.

ISBN: 979-8230842507

Written by Jihad Saffouri.

"In recognition of the individuals who have shaped my intellectual and personal growth: my family's devotion, my teachers' instruction, my friends' solidarity, and my colleagues' partnership. This work is indebted to them all."

"Words are but the shadows of actions; yet, in their silence, they shape worlds."

Preface

Our planet stands at a pivotal moment in its history. We are confronted by challenges like climate change, pollution, loss of biodiversity, and the depletion of vital natural resources—issues that are unlike any we have faced before. These crises not only jeopardize the environment but also threaten the stability of our societies, economies, and the future of life on Earth. Merely acknowledging these problems is no longer sufficient; urgent action is essential. *"Saving Our Planet: A Call to Action for a Sustainable Future"* is an urgent appeal for all of us—individuals, businesses, governments, and communities—to unite and take bold steps toward creating a sustainable world.

This book was written with a singular goal: to ignite change. As an environmental advocate, I have spent many years exploring the complexities of humanity's interaction with nature, working alongside scientists, policymakers, and communities to better understand the underlying causes of environmental degradation, and most importantly, to uncover the solutions that can help us restore balance. I have come to understand that change starts with awareness, but it is only sustained through decisive action.

Within these pages, I will take you through the most pressing environmental crises of our time—from the escalating climate emergency to the rapid extinction of species. But more importantly, I will demonstrate that these challenges are not insurmountable. With innovative solutions, strategic policy reforms, technological breakthroughs, and, above all, individual accountability, we can create a future in which both humanity and nature can flourish together.

The book is organized around key themes that examine various facets of sustainability, including renewable energy, waste management, biodiversity preservation, and the crucial roles of education and community engagement. Each chapter draws upon real-world examples and case studies that showcase successful initiatives from around the globe—stories that highlight the incredible power of human creativity and collective determination. These success stories offer a roadmap for what is achievable when we proactively address the environmental issues threatening our planet.

One of the central messages of this book is that sustainability is not a distant concept reserved for experts or policymakers. It is a principle that can and should be integrated into our everyday lives, whether through making sustainable choices, advocating for effective policy reforms, or supporting businesses and initiatives that prioritize the health of our planet. While the challenges we face are significant, the solutions are within our reach—and they require all of us to get involved.

I will also underscore the importance of collaboration. No single individual, organization, or nation can overcome the climate crisis alone. We must come together—across boundaries, industries, and communities—if we are to secure a sustainable future. The urgency of our environmental crisis calls for swift, bold action, but it also demands optimism and hope. Together, we have the power to protect and restore the Earth for the benefit of generations to come.

As you read this book, I encourage you to reflect on the role you can play in the movement to preserve our planet. Let these words inspire you, reminding you that even the smallest of actions can lead to meaningful change. This is not just a call to save the Earth; it is a call to save ourselves. The time to act is now.

Understanding our planet's crisis

Introduction: Saving Our Planet

The author of *Saving Our Planet: A Call to Action for a Sustainable Future* is an experienced environmental advocate and thought leader with a profound commitment to sustainability and climate action. With a strong foundation in environmental science and policy, the author has devoted much of their career to advancing solutions for some of the most pressing global challenges, including climate change, the loss of biodiversity, and the depletion of natural resources. This book is the result of years of research, collaboration with experts, and hands-on efforts to promote sustainable practices across communities, businesses, and governments.

Over the years, the author has worked alongside policymakers, businesses, and NGOs, advocating for environmental justice and the integration of green technologies. With a vision for a cleaner, more equitable world, the author underscores the importance of every individual's contribution to sustainability and the mitigation of climate change. Their work highlights the complex relationships between human actions, ecosystems, and the pressing need to restore balance to our planet's natural resources.

The author's work goes beyond theory, providing practical, real-world solutions. Through extensive travel and direct engagement with environmental leaders and local communities, the author has witnessed firsthand the devastating consequences of environmental degradation. These experiences have reinforced their strong commitment to environmental stewardship and have shaped the insights shared in *Saving Our Planet*.

A key aspect of the book is the author's thorough exploration of global environmental policies and their role in fostering sustainability. By examining crucial issues such as renewable energy, waste management, and conservation, the author offers readers tangible steps for creating change. The book also emphasizes the need for collective action, stressing that individuals, organizations, and governments all have a part to play in addressing ongoing environmental crises.

Saving Our Planet is more than just a book; it is a call for action. Through compelling examples, case studies, and actionable strategies, the author motivates readers to take responsibility for the planet's future. The book is written in an accessible and engaging style, designed to inspire people from all backgrounds to recognize the importance of sustainability and act with urgency. Whether through small daily actions or larger systemic changes, the author believes humanity can reverse the damage done and create a sustainable world for future generations.

In addition to their expertise in environmental matters, the author is a passionate educator, frequently speaking at events and workshops aimed at raising environmental awareness. Their work has garnered recognition from international environmental organizations and has played a significant role in the global movement toward a sustainable future.

Ultimately, *Saving Our Planet* reflects the author's belief in the power of informed, collective action. It serves as both a roadmap and an inspiration for anyone eager to contribute to the fight for a thriving, sustainable planet.

The global environmental crisis encompasses a range of interconnected challenges impacting the Earth's ecosystems and the life they support, including climate change, biodiversity loss, pollution, and resource depletion. These environmental issues have escalated to critical levels and are largely a consequence of human activities, requiring immediate attention and decisive action.

Climate change, driven by greenhouse gas emissions from fossil fuel use, is leading to rising global temperatures, more frequent extreme weather events, rising sea levels, and disruptions to agricultural systems (IPCC, 2021). The loss of biodiversity, caused mainly by deforestation, habitat destruction, and overexploitation of species, threatens numerous species and disrupts ecosystems essential to human survival, providing food, water, and medicine. Pollution, including contamination of air, water, and soil, continues to harm ecosystems and contribute to health problems for both humans and wildlife (UNEP, 2019). Furthermore, the depletion of natural resources, particularly non-renewable ones such as fossil fuels and minerals, raises concerns about the sustainability of human civilization (Rockström et al., 2009).

The urgency of addressing these environmental issues is undeniable, as their combined effects are already felt globally. Without prompt action, the world will face irreversible environmental damage, biodiversity loss, and severe economic and social consequences. Global cooperation and sustainable practices are essential to mitigate these risks and ensure a livable planet for future generations.

The historical roots of environmental degradation can be traced back to human development and industrialization. Early civilizations had localized environmental impacts, such as deforestation and soil depletion, but the scale of degradation dramatically increased with the advent of the Industrial Revolution in the 18th and 19th centuries. This period marked the beginning of mass production, urbanization, and increased reliance on fossil fuels, all of which contributed to significant environmental changes.

The Industrial Revolution, which began in Britain, led to the large-scale use of coal and the development of factories, steam engines, and transportation systems. These innovations spurred economic growth but also resulted in the increased release of pollutants into the atmosphere and waterways, deforestation, and the exploitation of natural resources at unprecedented rates. The environmental consequences included the degradation of air quality, widespread soil erosion, and the depletion of wildlife populations (Steffen et al., 2011).

As industrialization progressed, early warnings about the environmental impacts of human activity emerged. Scientists and environmental activists began to recognize the long-term consequences of unchecked development. In the late 19th and early 20th centuries, figures such as John Muir and Rachel Carson advocated for environmental conservation and awareness. Muir's work in the United States led to the establishment of national parks, and Carson's groundbreaking book *Silent Spring* (1962) highlighted the dangers of pesticide use, contributing to the rise of the modern environmental movement. These early efforts laid the foundation for future environmental advocacy and policy changes.

The historical roots of environmental degradation are thus deeply tied to industrial progress, with early warnings from scientific and activist communities signaling the need for sustainable development and environmental protection.

Climate change refers to long-term changes in temperature, precipitation patterns, and other atmospheric conditions on Earth. It is primarily driven by human activities that release greenhouse gases (GHGs) into the atmosphere, trapping heat and leading to global warming. Climate change is not just an environmental issue; it has far-reaching impacts on ecosystems, human health, and economies.

The primary drivers of climate change are the burning of fossil fuels, such as coal, oil, and natural gas, for energy and transportation. These activities release carbon dioxide (CO2) and other GHGs, which accumulate in the atmosphere and enhance the greenhouse effect. Deforestation, which reduces the Earth's capacity to absorb CO2 through trees, is another significant contributor. As forests are cleared for agriculture, urbanization, and industrial activities, large amounts of stored carbon are released, further exacerbating global warming.

The global rise in temperature is one of the most apparent effects of climate change. Since the late 19th century, average global temperatures have risen by approximately 1.1°C. This increase may seem small, but it has already led to significant changes in weather patterns. Extreme weather events, including more frequent and intense hurricanes, heatwaves, droughts, and heavy rainfall, are becoming more common as a result. For example, the frequency of heatwaves in Europe, the intensity of hurricanes in the Atlantic, and the severity of droughts in parts of Africa and the Americas have all been linked to climate change.

These changes in temperature and weather patterns have profound implications for agriculture, water resources, human health, and biodiversity. Addressing the root causes of climate change, particularly the reduction of fossil fuel use and the protection of forests, is essential to mitigating these risks and avoiding even more severe consequences in the future.

Climate change has profound effects on ecosystems and biodiversity, altering the delicate balance of life on Earth. Rising temperatures, shifting weather patterns, and other climate-related changes are disrupting habitats and threatening species with extinction. These shifts not only harm the environment but also pose significant risks to human survival, as ecosystems provide essential services such as food, water, and climate regulation.

One of the most visible effects of climate change on ecosystems is the rise in sea levels. Melting polar ice caps and glaciers, combined with the thermal expansion of seawater due to higher temperatures, contribute to rising sea levels. This leads to the flooding of coastal ecosystems, including mangroves, wetlands, and coral reefs, which are crucial habitats for numerous species. Coral reefs, in particular, are highly vulnerable to warming waters and ocean acidification, resulting in coral bleaching and the loss of biodiversity in these critical marine environments

Changing weather patterns also have significant impacts on ecosystems. For instance, altered rainfall patterns can lead to droughts in some regions and floods in others, both of which disrupt food and water supplies. Increased temperatures may also shift the geographic ranges of species, forcing them to migrate to new areas or adapt to new conditions. Species that cannot adapt quickly enough face a higher risk of extinction. The loss of biodiversity, in turn, diminishes ecosystem resilience, making it harder for ecosystems to recover from disturbances and providing fewer services to humans (Sala et al., 2000).

The consequences of ecosystem disruption extend far beyond environmental impacts. Healthy ecosystems provide vital services such as pollination, water purification, and climate regulation. For example, forests act as carbon sinks, absorbing CO2 from the atmosphere and mitigating the effects of climate change. If ecosystems are damaged or destroyed, these services are compromised, which can have devastating effects on agriculture, human health, and overall well-being. Additionally, the loss of biodiversity can lead to the collapse of food webs, affecting everything from crop production to the availability of seafood.

The disruption of ecosystems due to climate change is not just an environmental issue—it is a critical challenge for human survival. Protecting and restoring ecosystems is vital to maintaining the stability of the planet's life-support systems and ensuring a sustainable future for all.

Pollution is a growing global concern, affecting air, water, soil, and even noise, all of which contribute to the degradation of ecosystems and pose significant threats to human health and wildlife. The proliferation of pollutants has reached alarming levels, with industrial activities, agriculture, and urbanization being major sources of contamination.

Air Pollution is one of the most widespread forms of pollution, primarily caused by the burning of fossil fuels, industrial emissions, and vehicle exhaust. It releases harmful substances such as particulate matter (PM), sulfur dioxide (SO2), nitrogen oxides (NOx), and carbon monoxide (CO) into the atmosphere. These pollutants can cause respiratory diseases, cardiovascular problems, and even premature death in humans. Air pollution is also detrimental to wildlife, with some species experiencing reduced reproductive success or habitat degradation due to poor air quality (World Health Organization [WHO], 2021).

Water Pollution occurs when harmful substances, such as chemicals, heavy metals, and plastics, contaminate water bodies like rivers, lakes, and oceans. Industrial discharge, agricultural runoff (including pesticides and fertilizers), and untreated sewage are significant contributors. Contaminated water can lead to a range of health issues, including gastrointestinal diseases, infections, and poisoning from heavy metals. Marine life is particularly at risk, with species ingesting or becoming entangled in plastics, leading to injury or death.

Soil Pollution is caused by the accumulation of toxic substances such as pesticides, heavy metals, and industrial waste in the soil. This contamination harms plant life by affecting soil fertility and disrupting nutrient cycles. In turn, this reduces agricultural productivity, leading to food insecurity. Soil pollution also poses risks to humans through the consumption of contaminated crops and the leaching of toxins into groundwater supplies.

Noise Pollution is often overlooked but is growing in urban areas due to traffic, industrial activities, and other human-made sounds. Persistent exposure to high noise levels can lead to hearing loss, stress, sleep disturbances, and cardiovascular problems. Wildlife, particularly in urban areas, also faces challenges from noise pollution, as it interferes with communication, navigation, and mating behaviors.

The global impact of pollution is evident in several major pollution crises. For example, the **Great Smog of London** in 1952 resulted from a combination of industrial emissions and weather conditions, leading to thousands of deaths due to respiratory diseases. The **Minamata Bay disaster** in Japan (1950s) was caused by the release of mercury into the bay by industrial activities, leading to severe mercury poisoning in both humans and wildlife. More recently, the widespread contamination of oceans by plastics, particularly in the **Great Pacific Garbage Patch**, has become a global environmental crisis.

Pollution continues to have devastating effects on human health, wildlife, and ecosystems, underscoring the need for comprehensive policies, technological innovations, and global cooperation to mitigate its impact.

Deforestation and land degradation are significant drivers of environmental decline, contributing to the loss of biodiversity, disruption of ecosystems, and challenges to agriculture and food security. These issues are interlinked, with deforestation often being a primary cause of land degradation and both processes exacerbating climate change.

Extent of Deforestation and Its Causes

Deforestation refers to the large-scale clearing or destruction of forests for various purposes, including agricultural expansion, logging, infrastructure development, and urbanization. According to the Food and Agriculture Organization (FAO), approximately 10 million hectares of forest are lost each year (FAO, 2020). This loss is particularly alarming in tropical regions, such as the Amazon rainforest, where deforestation rates have been driven by agricultural activities like cattle ranching and palm oil production, as well as illegal logging and mining. The conversion of forests into agricultural land and urban areas disrupts ecosystems, alters local climates, and accelerates soil erosion.

Link Between Deforestation and Loss of Biodiversity

Forests are home to a vast majority of the world's terrestrial species, making them crucial to maintaining biodiversity. Deforestation leads to the fragmentation of habitats, causing species to lose their homes and food sources. As forests are destroyed or fragmented, many species are pushed toward extinction due to the inability to adapt or migrate to new habitats. For example, the deforestation of the Amazon basin has significantly threatened species such as the jaguar, the giant river otter, and various bird species. The loss of biodiversity also disrupts ecosystem services such as pollination, water purification, and climate regulation, which are vital for both wildlife and human populations (Pimm et al., 2014).

Consequences of Land Degradation on Agriculture and Food Security

Land degradation refers to the deterioration of the land's productive capacity, often caused by human activities such as deforestation, overgrazing, and the use of harmful agricultural practices. This degradation manifests in the form of soil erosion, desertification, reduced soil fertility, and the loss of arable land. As soil quality declines, agricultural productivity suffers, leading to lower crop yields and reduced food availability. In regions where land degradation is widespread, such as parts of sub-Saharan Africa and South Asia, food security becomes a major concern, with millions of people facing hunger and malnutrition due to declining agricultural output.

Furthermore, land degradation exacerbates the impacts of climate change. Reduced vegetation cover and poor soil health contribute to increased carbon emissions, while the loss of forests and fertile land makes it harder for communities to adapt to changing climate conditions. The combination of deforestation, land degradation, and climate change poses a serious threat to global food systems, making it essential to adopt sustainable land management practices and policies aimed at protecting ecosystems and restoring degraded lands.

Resource depletion and overconsumption are key factors driving environmental degradation. The unsustainable use of natural resources has intensified with population growth, industrialization, and increasing demands for goods and services, pushing the planet's ecosystems beyond their capacity to regenerate.

Unsustainable Use of Natural Resources

Natural resources, such as water, minerals, forests, and fossil fuels, are essential for human survival and economic development. However, the rate at which these resources are extracted and used is far exceeding the Earth's regenerative capacity. Overconsumption and inefficient resource use, especially in high-income countries, contribute to the depletion of these resources and their subsequent inability to support future generations. Unsustainable agricultural practices, industrial activities, and urbanization continue to put immense pressure on the planet's resources, leading to habitat destruction, pollution, and the degradation of essential ecosystems.

Overfishing, Overuse of Freshwater, and Fossil Fuels

One of the most concerning aspects of resource depletion is **overfishing**, which threatens marine biodiversity and the livelihoods of communities dependent on fish as a primary food source. Many fish populations are being harvested faster than they can reproduce, leading to the collapse of entire fisheries. The **overuse of freshwater** is another critical issue. With increasing global demand for water in agriculture, industry, and domestic consumption, many regions are experiencing water scarcity. The depletion of groundwater reserves and the contamination of freshwater bodies further exacerbate this problem, threatening both human populations and ecosystems (Postel, 1999).

The **extraction and burning of fossil fuels**—coal, oil, and natural gas—remain primary contributors to both resource depletion and climate change. Fossil fuel consumption has increased rapidly over the past century, with the resulting carbon emissions significantly impacting global temperatures and air quality. The continued reliance on these non-renewable resources accelerates their depletion, leading to energy crises and further environmental damage. As reserves become harder to access, the environmental impact of extraction, such as oil spills, habitat destruction, and air pollution, worsens.

Effects of Overconsumption on the Planet's Ability to Regenerate

Overconsumption places immense strain on the Earth's regenerative capacity. Ecosystems, such as forests, wetlands, and oceans, play a crucial role in absorbing carbon, recycling nutrients, and providing food and shelter for countless species. However, human demands for resources exceed the natural regeneration rates of these ecosystems, leading to habitat destruction and the degradation of biodiversity. As a result, the planet's ability to support life and recover from environmental stress is increasingly compromised.

This unsustainable use of resources has far-reaching consequences for human societies, from food and water insecurity to loss of biodiversity and worsening climate change. Moving toward sustainable consumption and production practices is essential to reducing pressure on the Earth's ecosystems and ensuring the planet can regenerate and support future generations.

Human population growth plays a significant role in environmental strain, amplifying the pressures on natural resources, ecosystems, and climate systems. As the global population continues to rise, the demand for food, water, energy, and land intensifies, leading to increased environmental degradation. Managing this growth is one of the most pressing challenges for ensuring a sustainable future for both humanity and the planet.

Relationship Between Population Growth and Environmental Strain

The relationship between population growth and environmental strain is primarily driven by the increased demand for resources and the greater production of waste. As more people inhabit the Earth, more land is needed for housing, agriculture, and infrastructure, leading to deforestation, habitat destruction, and a loss of biodiversity. Additionally, the growing population contributes to higher consumption levels, further depleting natural resources such as water, fossil fuels, and minerals. This leads to a cycle of overuse, where the Earth's capacity to regenerate is surpassed by human demand, resulting in environmental degradation and a depletion of vital resources (Cohen, 2006).

Urbanization, Waste, and Energy Demands

Urbanization is one of the key drivers of environmental strain associated with population growth. As more people migrate to cities, urban areas expand, often at the expense of natural habitats. This expansion leads to increased waste production, air and water pollution, and greater demand for energy. Cities are responsible for a significant portion of global greenhouse gas emissions due to transportation, industrial activities, and energy consumption. The growing need for electricity, heating, and cooling in urban centers also contributes to the depletion of fossil fuels, further exacerbating climate change.

Waste generation in urban areas has become a critical environmental issue, with much of it going to landfills or being incinerated, both of which cause environmental pollution. Plastic waste, for example, has accumulated in landfills and oceans, posing significant threats to marine and terrestrial wildlife. Urban areas also place immense pressure on waste management systems, leading to inefficiencies and further contamination of the environment.

Energy demand is another area where population growth has a significant impact. As the population increases, so does the need for energy to power homes, industries, and transportation systems. This growing demand often relies on non-renewable energy sources, such as coal, oil, and natural gas, which contribute to greenhouse gas emissions and climate change. Transitioning to renewable energy sources is crucial to reducing the environmental impact of population growth.

The Challenge of Managing a Growing Global Population

Managing a growing global population is a complex challenge that requires addressing not only population control but also the sustainable use of resources, equitable access to goods and services, and the reduction of environmental impact. Balancing population growth with the planet's ecological limits demands significant efforts in areas such as family planning, education, and poverty reduction, as well as policy measures aimed at promoting sustainable lifestyles and reducing consumption. Technological advancements, such as renewable energy and sustainable agriculture, are also critical in mitigating the effects of population growth on the environment (Sachs, 2015).

In conclusion, while population growth is a key factor contributing to environmental strain, the way humanity chooses to manage this growth—through sustainable practices, resource management, and technological innovation—will determine the future of the planet. The challenge lies in meeting the needs of a growing population without compromising the ability of future generations to meet their own needs.

Global politics and economic systems play a significant role in driving the environmental crisis, often prioritizing short-term profits over long-term sustainability. The intricate relationship between political decisions, economic structures, and environmental degradation is a key factor in the continued deterioration of ecosystems, climate stability, and resource availability. These systems often reinforce unsustainable practices, contributing to pollution, deforestation, and overconsumption.

Global Politics and Economic Systems

The structure of the global economy, characterized by industrialization, consumerism, and economic growth, is closely tied to environmental damage. Political decisions made by governments, often influenced by powerful economic interests, frequently prioritize economic development and growth over environmental protection. For instance, policies that support fossil fuel subsidies or deregulate industries can lead to increased environmental pollution and unsustainable resource extraction. The globalization of trade and the pursuit of economic growth in developing countries further intensify resource use, as industries are incentivized to exploit cheap labor and natural resources without considering environmental sustainability.

The global political landscape also contributes to the crisis through international trade agreements and the lack of enforceable environmental regulations. Economic globalization has often led to environmental degradation, as multinational corporations exploit natural resources in countries with lax environmental laws. While some countries have made strides in environmental policy, the lack of global cooperation on issues like climate change, pollution, and biodiversity loss often hinders significant progress.

Short-Term Profit Motives vs. Long-Term Sustainability

A major driver of the environmental crisis is the short-term profit motive prevalent in both corporate and governmental decision-making. The pursuit of immediate economic benefits, such as increased production, resource extraction, or market expansion, often leads to environmental harm. For example, industries such as oil, mining, and agriculture focus on maximizing profits in the short term, often disregarding the long-term environmental consequences of their actions. This approach is exemplified by the extraction of fossil fuels, which not only contributes to climate change but also pollutes land, air, and water.

Moreover, in many instances, governments prioritize short-term economic gains to secure political and economic power, sometimes at the expense of sustainable development. Environmental regulations are often weakened or ignored to attract foreign investment or boost economic growth, creating a vicious cycle of environmental degradation. In this context, there is a tension between policies that encourage economic growth and those that protect the environment, making it difficult to achieve long-term sustainability.

Examples of Policies and Economic Activities Exacerbating Environmental Damage

Several policies and economic activities have exacerbated environmental damage. One example is the **subsidization of fossil fuels**, which incentivizes their continued extraction and consumption, contributing significantly to greenhouse gas emissions and global warming. For decades, countries like the United States and Saudi Arabia have provided subsidies to the fossil fuel industry, keeping prices artificially low and encouraging overuse (Coady et al., 2019).

Another example is **deforestation** driven by agricultural policies that encourage land conversion for crops like palm oil, soy, and cattle ranching. In countries such as Brazil and Indonesia, economic policies have led to massive deforestation, contributing to habitat loss, biodiversity decline, and increased carbon emissions (Boucher et al., 2018). Similarly, the **overexploitation of marine resources** through policies that favor overfishing has led to the collapse of fish stocks in many parts of the world, impacting both marine ecosystems and the livelihoods of millions of people.

Finally, **industrialization and urbanization** policies that prioritize economic development have led to environmental harm, including air and water pollution, as well as the loss of green spaces. For instance, rapid urbanization in developing countries often involves the clearing of forests or wetlands to make way for housing and infrastructure, leading to soil erosion, reduced biodiversity, and increased carbon emissions.

The tension between economic growth and environmental sustainability is a significant barrier to addressing the environmental crisis. Shifting toward policies that promote sustainable development and long-term ecological balance will require a fundamental rethinking of global economic priorities.

Environmental degradation disproportionately impacts poorer nations, where communities are more vulnerable to the effects of climate change, pollution, and resource depletion. These countries often lack the infrastructure, resources, and political power to effectively mitigate or adapt to environmental challenges. As a result, the burden of environmental harm falls most heavily on the world's poorest populations, exacerbating social inequality and creating pressing human rights concerns.

Disproportionate Impact of Environmental Degradation on Poorer Nations

The poorest nations are often located in regions that are most vulnerable to the effects of environmental degradation, such as low-lying coastal areas, drought-prone regions, and areas with fragile ecosystems. These countries typically have fewer financial and technological resources to cope with the consequences of climate change, such as extreme weather events, rising sea levels, and droughts. For example, small island nations in the Pacific, such as the Maldives and Kiribati, face existential threats from sea level rise, while Sub-Saharan Africa is particularly affected by desertification and water scarcity. These countries often rely heavily on agriculture, which is increasingly compromised by changing weather patterns and land degradation.

Additionally, many poorer nations are heavily dependent on natural resources for their economic livelihoods, such as agriculture, fishing, and forestry. Environmental degradation, such as soil erosion, deforestation, and overfishing, undermines these industries, leading to loss of income, food insecurity, and rising poverty levels.

The inability to protect these industries further deepens the vulnerability of these nations to global environmental crises.

Environmental Justice and the Need for Equitable Solutions

Environmental justice advocates for equitable solutions to environmental problems, recognizing that the impacts of environmental degradation are not shared equally. Those who contribute least to environmental harm, typically poorer nations and marginalized communities, are often the ones who suffer the most. This injustice is particularly evident in the case of climate change, where industrialized nations, which have historically been the largest polluters, continue to emit large amounts of greenhouse gases, while poorer nations bear the brunt of climate impacts.

Efforts to address environmental justice must focus on ensuring that vulnerable populations have access to resources and opportunities to protect their environment and adapt to climate change. This includes providing financial support for adaptation and mitigation projects, promoting sustainable development, and ensuring that the voices of marginalized communities are heard in environmental decision-making. Policies that promote equitable distribution of resources, strengthen community resilience, and reduce environmental risks are essential to achieving environmental justice (Bullard, 2005).

Climate Refugees and Human Rights Concerns

Climate change is creating a growing number of **climate refugees**, people who are forced to migrate due to the impacts of environmental degradation, such as extreme weather events, sea level rise, and resource scarcity. According to the United Nations, millions of people are already displaced by climate-related disasters, and this number is expected to rise significantly in the coming decades (IOM, 2020). These refugees face a range of challenges, including loss of livelihood, lack of access to basic services, and threats to their safety and dignity.

The rights of climate refugees are a growing concern. Unlike traditional refugees who are protected under international law, climate refugees are not yet recognized as a distinct category under international law, leaving them vulnerable to exploitation and denial of rights. Ensuring the protection and rights of climate refugees is an urgent human rights issue that requires international cooperation and legal frameworks to address the unique challenges they face. This includes creating policies that ensure safe migration, provide access to essential services, and offer legal protections for those displaced by environmental factors (McAdam, 2012).

In conclusion, the environmental crisis exacerbates global inequality, with the poorest nations bearing the brunt of environmental harm. Achieving environmental justice requires not only addressing the root causes of environmental degradation but also ensuring that solutions are equitable and inclusive, protecting the rights of those most affected by the crisis, including climate refugees.

Global efforts to combat the environmental crisis involve a combination of international agreements, global organizations, and grassroots movements. These initiatives aim to address the pressing issues of climate change, pollution, biodiversity loss, and resource depletion. While there have been notable successes in some areas, ongoing challenges remain in achieving widespread and effective action to mitigate the crisis.

The Role of International Agreements, such as the Paris Agreement

International agreements play a critical role in coordinating global efforts to address the environmental crisis, particularly in the context of climate change. The **Paris Agreement**, adopted in 2015, is one of the most significant international climate accords. Its goal is to limit global warming to well below 2°C above pre-industrial levels, with an aim of limiting it to 1.5°C. The agreement sets targets for reducing greenhouse gas emissions, with each country pledging to submit nationally determined contributions (NDCs) outlining their climate action plans.

The Paris Agreement is groundbreaking in its inclusivity, as it brings together both developed and developing nations to address climate change collectively. It also emphasizes financial support for developing countries to help them transition to sustainable development and adapt to climate impacts. While the agreement has been praised for its global approach and its focus on long-term sustainability, challenges remain in ensuring that all countries meet their commitments and that the actions taken are sufficient to limit warming to safe levels (UNFCCC, 2015).

The Impact of Organizations like the United Nations and Greenpeace

Organizations like the **United Nations (UN)** and **Greenpeace** play pivotal roles in raising awareness, advocating for policy changes, and coordinating global efforts to combat environmental degradation.

The **United Nations** has been at the forefront of international environmental governance through initiatives such as the **UN Environment Programme (UNEP)** and the **UN Framework Convention on Climate Change (UNFCCC)**. The UN facilitates global dialogues on environmental issues, such as the annual **COP (Conference of the Parties)** meetings, where countries come together to negotiate climate policies and share progress on environmental goals. The UN also promotes the **Sustainable Development Goals (SDGs)**, with Goal 13 specifically focused on taking urgent action to combat climate change and its impacts.

Greenpeace, as a prominent environmental organization, has been instrumental in raising public awareness and holding governments and corporations accountable for environmental harm. Through campaigns and direct action, Greenpeace has pushed for stronger environmental protection laws, the cessation of harmful practices such as deforestation and overfishing, and the transition to renewable energy sources. Greenpeace's advocacy for policies that promote sustainability and environmental justice has made it a key player in global environmental activism.

Success Stories and Ongoing Challenges

While global efforts have led to some significant successes, there are ongoing challenges in fully addressing the environmental crisis.

Success Stories

- The **Montreal Protocol**, signed in 1987, is one of the most successful international agreements, as it led to the global phase-out of ozone-depleting substances. This agreement has had a major positive impact on the recovery of the ozone layer and is considered a model for future environmental agreements (UNEP, 2017).
- The **growth of renewable energy** has been another success story, with significant increases in the use of solar, wind, and other clean energy sources. Countries like Denmark and Germany have made substantial progress in transitioning to renewable energy, showing that it is possible to decouple economic growth from fossil fuel dependence.
- **Forest conservation efforts**, such as the establishment of protected areas and the implementation of sustainable logging practices, have helped preserve biodiversity and prevent deforestation in several regions, including the Amazon and Southeast Asia (WWF, 2020).

Ongoing Challenges

Despite these successes, there are significant hurdles that continue to impede progress in combating the environmental crisis.

- **Political and economic resistance** from fossil fuel industries, along with the short-term profit motives of governments and corporations, often slow down the implementation of meaningful environmental policies.
- **Funding and resources** for climate change mitigation and adaptation remain insufficient, particularly for vulnerable countries in the Global South that are facing the most severe impacts of climate change.
- **Biodiversity loss** continues at an alarming rate, with many species going extinct due to habitat destruction, pollution, and climate change, despite efforts to establish protected areas and promote conservation.

In conclusion, while there have been notable international efforts and success stories in combating the environmental crisis, significant challenges remain. Global cooperation, stronger political will, and continued advocacy are essential to achieving long-term sustainability and addressing the pressing environmental issues of our time.

The transition from fossil fuels to renewable energy sources and the adoption of technological innovations are essential to mitigating the environmental crisis. Renewable energy and green technologies offer the promise of reducing greenhouse gas emissions, lowering our dependence on finite resources, and creating a sustainable energy future. However, while these technologies hold significant potential, they also come with limitations that must be addressed to ensure their widespread adoption and effectiveness.

The Shift from Fossil Fuels to Renewable Energy Sources

The shift from fossil fuels (coal, oil, and natural gas) to renewable energy sources is a critical strategy for addressing climate change and reducing environmental harm. Fossil fuel combustion is the primary source of greenhouse gas emissions, which are driving global warming and climate-related disasters. In contrast, renewable energy sources such as **solar**, **wind**, and **geothermal** produce little to no carbon emissions and can provide a sustainable supply of energy.

- **Solar Energy**: Solar power harnesses the energy of the sun through photovoltaic panels or solar thermal systems. It is abundant, renewable, and rapidly growing in adoption worldwide. As technology improves and costs decrease, solar energy is becoming increasingly competitive with traditional energy sources. Countries like China, India, and the United States are investing heavily in solar infrastructure to meet their energy demands while reducing carbon footprints (IRENA, 2020).
- **Wind Energy**: Wind power involves using turbines to convert the kinetic energy of wind into electricity. It is a rapidly expanding source of renewable energy, with offshore wind farms gaining prominence in regions with strong wind resources. Wind energy has the potential to provide a substantial portion of the world's energy needs, particularly in areas with favorable wind conditions like Europe and North America.
- **Geothermal Energy**: Geothermal energy taps into the heat stored beneath the Earth's surface to generate electricity or provide direct heating. It is a reliable and consistent energy source, with significant potential in areas near tectonic plate boundaries, such as Iceland, the Philippines, and parts of the United States. Although geothermal energy is still less widespread compared to solar and wind, it offers a consistent, low-emission alternative for base-load power generation.

Technological Solutions: Electric Vehicles and Carbon Capture

In addition to renewable energy, technological innovations like **electric vehicles (EVs)** and **carbon capture and storage (CCS)** offer promising solutions to reduce the environmental impact of human activities.

- **Electric Vehicles (EVs)**: EVs offer a sustainable alternative to traditional gasoline-powered vehicles by reducing emissions from the transportation sector, which is a significant contributor to global greenhouse gas emissions. The transition to EVs is gaining momentum, supported by advancements in battery technology, falling prices, and the expansion of charging infrastructure. Governments around the world are offering incentives to promote EV adoption, with countries like Norway and the Netherlands leading the way in electric vehicle market share. However, challenges remain in scaling up EV production, improving battery technology, and addressing the environmental impact of battery disposal and mining for raw materials.
- **Carbon Capture and Storage (CCS)**: CCS is a technology that captures carbon dioxide emissions from industrial processes and power plants and stores it underground, preventing it from entering the atmosphere. This technology has the potential to reduce emissions from sectors that are difficult to decarbonize, such as heavy industry and fossil fuel-based energy generation. However, CCS is still in the early stages of development, and its widespread deployment faces challenges related to cost, scalability, and the long-term storage of carbon.

The Promise and Limitations of Green Technology

Green technologies, including renewable energy sources and innovations like EVs and CCS, offer great promise in addressing the environmental crisis. They provide a path toward a low-carbon, sustainable future and can contribute significantly to mitigating climate change. However, these technologies also come with limitations that need to be addressed for them to reach their full potential.

- **Intermittency and Storage**: Renewable energy sources like solar and wind are intermittent, meaning they do not produce energy consistently. This can lead to challenges in ensuring a reliable power supply, especially during periods of low sunlight or wind. Energy storage solutions, such as batteries, are critical to overcoming this challenge, but large-scale storage technology is still expensive and not yet sufficiently developed to meet global needs.
- **Environmental and Resource Impacts**: While renewable energy technologies are generally cleaner than fossil fuels, they are not entirely free from environmental impacts. For example, the mining of raw materials for solar panels, wind turbines, and electric vehicle batteries can have significant environmental consequences, including habitat destruction, pollution, and resource depletion. Additionally, the disposal and recycling of these technologies at the end of their life cycle present environmental challenges that must be addressed (Gunter et al., 2019).
- **Equity and Access**: Green technologies must be accessible and affordable to all populations, especially those in developing countries that are most vulnerable to climate change. The transition to renewable energy and the

adoption of green technologies requires significant investment, and many low-income countries may struggle to access the necessary resources or infrastructure. International cooperation, financial support, and technology transfer are essential to ensure that all nations can benefit from these technologies.

In conclusion, while renewable energy and technological innovations offer a pathway to a more sustainable future, overcoming the limitations of these technologies and addressing the challenges they present will be crucial in mitigating the environmental crisis. Continued research, investment, and global cooperation will be essential in ensuring the success of green technologies and their role in creating a sustainable, low-carbon world.

Agriculture plays a central role in both contributing to and mitigating the environmental crisis. Current agricultural practices, while essential for feeding the global population, are major drivers of environmental degradation, including climate change, deforestation, and biodiversity loss. The shift toward **sustainable agriculture** and **food systems** offers a solution to reduce these impacts while ensuring food security for future generations.

The Environmental Impact of Current Agricultural Practices

Traditional agricultural practices, especially those employed in large-scale industrial farming, have significant negative effects on the environment. Key environmental impacts of current agriculture include:

- **Greenhouse Gas Emissions**: Agriculture is responsible for approximately 25% of global greenhouse gas emissions, largely due to livestock production (methane emissions), the use of synthetic fertilizers (nitrous oxide emissions), and energy-intensive farming practices. Livestock farming,

particularly cattle, contributes significantly to methane emissions, a potent greenhouse gas with a much higher warming potential than carbon dioxide.
- **Deforestation**: Large-scale agricultural expansion, particularly for crops like soy and palm oil and for livestock grazing, has led to significant deforestation, particularly in tropical regions such as the Amazon. Deforestation not only contributes to carbon emissions but also destroys critical biodiversity and disrupts ecosystems that provide essential services.
- **Water Use and Pollution**: Agriculture is a leading cause of water use and pollution worldwide. Irrigation for crops depletes freshwater resources, particularly in arid regions, while agricultural runoff containing pesticides, fertilizers, and sediment often contaminates water bodies, harming aquatic ecosystems and polluting drinking water sources.

The Role of Sustainable Agriculture in Mitigating Climate Change

Sustainable agriculture aims to reduce the negative environmental impacts of farming while maintaining or enhancing agricultural productivity. By adopting practices that promote environmental stewardship, sustainable agriculture can play a crucial role in mitigating climate change and enhancing food security.

- **Carbon Sequestration**: Practices such as **no-till farming**, **cover cropping**, and **agroforestry** can help sequester carbon in the soil, reducing the amount of carbon dioxide in the atmosphere. By improving soil health and increasing organic matter content, these practices enhance soil's ability to store carbon, potentially offsetting a significant portion of agricultural emissions.

- **Reduced Chemical Inputs**: Sustainable agriculture encourages the reduction of synthetic fertilizers and pesticides, which can harm soil health, water quality, and biodiversity. Practices like **integrated pest management (IPM)** and **organic farming** promote natural alternatives to chemicals, which not only protect the environment but can also improve long-term soil fertility.
- **Agroecology and Permaculture**: Agroecological practices focus on designing farming systems that mimic natural ecosystems. This includes crop diversification, reduced reliance on external inputs, and greater emphasis on local knowledge and biodiversity conservation. These practices can make farms more resilient to climate change, reduce environmental harm, and promote social equity.

Innovations in Food Systems: Plant-Based Diets and Regenerative Farming

Beyond farming practices, innovations in food systems—such as promoting plant-based diets and **regenerative farming**—hold the potential to transform agriculture and reduce its environmental footprint.

- **Plant-Based Diets**: Shifting toward plant-based diets is one of the most effective ways to reduce the environmental impact of food production. Plant-based foods generally require fewer natural resources (land, water, and energy) and produce lower greenhouse gas emissions compared to animal-based products. Studies show that if global diets shifted away from meat and dairy toward plant-based foods, global greenhouse gas emissions from food production could be significantly reduced. Promoting plant-based diets, particularly in high-income countries, could play a crucial role in reducing the environmental footprint of food systems.

- **Regenerative Farming**: Regenerative farming goes beyond sustainable agriculture by focusing on restoring and revitalizing soil health and ecosystems. Practices such as **rotational grazing**, **composting**, and **agroforestry** are designed to improve soil biodiversity, enhance water retention, and increase carbon sequestration. Regenerative agriculture aims to restore the land to a more natural, resilient state, and can be particularly effective in reversing soil degradation, which is a significant issue globally. It also offers the potential for higher food production with lower environmental costs in the long term.

Conclusion: The Path to Sustainable Agriculture and Food Systems

The transition to sustainable agriculture and food systems is essential for addressing the environmental crisis and ensuring food security in the future. While the environmental impact of current agricultural practices is significant, innovative solutions like plant-based diets, regenerative farming, and carbon-sequestering practices offer promising pathways for reducing emissions, conserving natural resources, and maintaining biodiversity. However, widespread adoption of these solutions will require policy changes, financial investments, and greater public awareness about the importance of sustainable food systems.

Conservation Efforts and Biodiversity Preservation

Biodiversity—the variety of life on Earth—is fundamental to the health of ecosystems, human well-being, and the resilience of the planet in the face of environmental challenges. Conservation efforts are critical in preserving biodiversity, protecting endangered species, and ensuring that ecosystems continue to provide essential services such as clean air, water, food, and climate regulation. Despite the ongoing threats to biodiversity, global conservation initiatives, protected areas, and successful wildlife protection and habitat restoration efforts offer hope for reversing the trends of species loss and environmental degradation.

The Importance of Preserving Biodiversity for Ecosystem Health

Biodiversity plays a vital role in maintaining the stability and function of ecosystems. Healthy ecosystems provide numerous services that support life on Earth, including:

- **Pollination**: Biodiversity supports pollination, which is essential for the reproduction of many plants, including agricultural crops. Without pollinators like bees, butterflies, and birds, food production would be severely affected, leading to significant challenges in feeding the growing global population.
- **Soil Fertility**: A diverse range of organisms, from microorganisms to larger animals, contribute to the health of soils by decomposing organic matter, recycling nutrients, and enhancing soil structure. Healthy soils are essential for agriculture and ecosystem functioning.
- **Climate Regulation**: Biodiversity helps regulate the climate through carbon sequestration, particularly in forests, wetlands, and oceans. The loss of biodiversity, such as the destruction of rainforests or the depletion of coral reefs, reduces the Earth's ability to absorb carbon, exacerbating climate change.
- **Resilience to Environmental Stress**: Biodiversity enhances the resilience of ecosystems to environmental stressors such as climate change, pollution, and habitat loss. A diverse array of species allows ecosystems to better

withstand and recover from disturbances, ensuring that essential ecosystem services continue to be provided.

Despite the critical importance of biodiversity, human activities have caused unprecedented rates of species loss and ecosystem degradation, often referred to as the **sixth mass extinction**. Habitat destruction, pollution, overexploitation, and climate change are driving the rapid decline of species, threatening ecosystem stability and resilience.

Global Conservation Initiatives and Protected Areas

To address the biodiversity crisis, numerous global conservation initiatives and protected area systems have been established. These initiatives focus on preserving habitats, protecting endangered species, and promoting sustainable development.

- **Protected Areas**: The establishment of protected areas, such as national parks, nature reserves, and marine protected areas (MPAs), is a cornerstone of global biodiversity conservation efforts. These areas are designated to prevent human activities like logging, mining, and urbanization that threaten ecosystems. Protected areas also serve as refuges for endangered species and can support the recovery of ecosystems.
- **The Convention on Biological Diversity (CBD)**: The **CBD**, adopted in 1992, is an international treaty aimed at conserving biodiversity, promoting sustainable use of natural resources, and ensuring the fair and equitable sharing of the benefits of biodiversity. The CBD has facilitated the creation of protected areas, biodiversity monitoring programs, and global efforts to prevent the loss of species and habitats.
- **The Global Environment Facility (GEF)**: The **GEF**

provides financial support for conservation projects around the world. It focuses on protecting biodiversity, preventing land degradation, conserving water resources, and mitigating climate change. Through its funding, the GEF has helped to establish protected areas, restore ecosystems, and support the sustainable management of natural resources (GEF, 2020).

- **International Union for Conservation of Nature (IUCN)**: The **IUCN** is a global organization that works to conserve biodiversity through research, policy advocacy, and support for conservation initiatives. The IUCN's **Red List of Threatened Species** is an important tool for monitoring the status of species worldwide and identifying those at risk of extinction. The IUCN also promotes the establishment of **Key Biodiversity Areas (KBAs)**, which are critical for the survival of threatened species and ecosystems. (IUCN, 2020).

Successes in Wildlife Protection and Habitat Restoration

While biodiversity loss remains a significant challenge, there have been notable successes in wildlife protection and habitat restoration that offer hope for the future.

- **Wildlife Protection**: Conservation efforts have led to the recovery of several iconic species that were once on the brink of extinction. The **bald eagle**, **gray wolf**, and **American bison** in North America have all made remarkable comebacks thanks to dedicated conservation programs, habitat restoration, and legal protections. In Africa, successful anti-poaching efforts and habitat conservation have contributed to the recovery of species such as the **black rhinoceros** and **mountain gorillas**.

- **Marine Conservation**: Marine conservation efforts, such as the establishment of marine protected areas (MPAs), have proven successful in protecting ocean ecosystems and wildlife. The creation of the **Great Barrier Reef Marine Park** in Australia has helped preserve the world's largest coral reef system. In the Pacific Ocean, the **Papahānaumokuākea Marine National Monument** in the United States is a large marine protected area that supports diverse marine life, including sea turtles and coral reefs.
- **Habitat Restoration**: Habitat restoration projects, including reforestation and wetland restoration, have demonstrated the potential for reversing environmental degradation. The **Atlantic Forest Restoration Pact** in Brazil is an ambitious effort to restore large areas of the endangered Atlantic Forest, which has been heavily impacted by deforestation. In the United Kingdom, the **rewilding** of former agricultural lands to create more natural habitats for species such as the **European wildcat** and **beaver** has shown that habitat restoration can support biodiversity recovery.

Conclusion: The Path to Biodiversity Preservation

Preserving biodiversity is essential for maintaining ecosystem health and ensuring a sustainable future for all life on Earth. Global conservation initiatives, the establishment of protected areas, and successful wildlife protection and habitat restoration projects offer hope for reversing the trends of biodiversity loss. However, continued efforts are needed to address the drivers of biodiversity decline, including habitat destruction, pollution, and climate change. By prioritizing biodiversity conservation and investing in sustainable practices, we can help ensure a healthy planet for future generations.

Waste Management and Circular Economy

The growing volume of waste produced globally is a significant environmental challenge. Improper waste management leads to pollution, resource depletion, and health hazards, making it essential to adopt more sustainable waste practices. The **circular economy** offers a solution by promoting waste reduction, recycling, and the sustainable use of resources. In this context, effective waste management strategies can help reduce environmental damage, conserve resources, and foster a more sustainable future.

The Global Waste Crisis and Its Environmental Consequences

Waste generation has increased dramatically in recent decades due to urbanization, population growth, and industrialization. According to the **World Bank (2018)**, the world generates over 2 billion tons of municipal solid waste each year, and this figure is expected to rise by 70% by 2050. This waste has several environmental consequences:

- **Pollution**: Improper waste disposal, especially in landfills and incinerators, leads to the release of toxic chemicals into the soil, water, and air. For example, plastic waste can end up in oceans, harming marine life and ecosystems. The accumulation of non-biodegradable materials like plastics in landfills is a significant contributor to environmental pollution.
- **Greenhouse Gas Emissions**: Decomposing organic waste in landfills produces methane, a potent greenhouse gas that contributes to climate change. Waste incineration also releases carbon dioxide and other pollutants into the atmosphere, exacerbating global warming.

- **Resource Depletion**: Disposing of valuable materials, such as metals, plastics, and textiles, in landfills instead of recycling or reusing them contributes to the depletion of natural resources. This unsustainable approach to resource use increases the demand for mining, fossil fuels, and raw materials, further exacerbating environmental degradation.

The Concept of a Circular Economy and Waste Reduction Strategies

The **circular economy** is a sustainable model that seeks to minimize waste and make the most of available resources. Unlike the traditional linear economy, which follows a "take, make, dispose" pattern, the circular economy promotes a "reduce, reuse, recycle" approach. It focuses on keeping materials in use for as long as possible and reducing the need for new resources.

Key principles of the circular economy include:

- **Design for Longevity**: Products are designed with longer lifespans and the possibility of reuse, repair, or recycling in mind. This reduces the amount of waste generated and conserves resources.
- **Recycling and Upcycling**: Rather than discarding products at the end of their life, recycling ensures that materials can be repurposed or reused in new products. Upcycling takes this a step further by turning waste materials into products of higher value.
- **Product-as-a-Service**: In the circular economy, businesses offer products as services rather than selling physical goods. For example, companies may lease products like furniture or electronics, allowing them to be returned, refurbished, and reused. This reduces the need for continuous production and minimizes waste.

- **Waste-to-Energy**: Some waste, particularly organic waste, can be converted into renewable energy through processes like anaerobic digestion or incineration. While not a perfect solution, this approach can help divert waste from landfills and generate energy.

Examples of Countries or Cities Leading in Waste Management

Several countries and cities around the world have adopted innovative waste management practices and circular economy principles to address the waste crisis. These examples demonstrate how effective waste management strategies can contribute to environmental sustainability.

- **Sweden**: Sweden has become a global leader in waste management and recycling. The country has developed a system where nearly 99% of waste is recycled, converted into energy, or reused. Sweden operates a highly efficient **waste-to-energy** program, where waste is incinerated to produce electricity and heating. Additionally, the country's robust recycling system ensures that materials like paper, plastic, and glass are consistently recycled, reducing landfill use. Sweden has also set ambitious goals for a **circular economy**, including reducing waste generation and promoting sustainable production and consumption practices (Swedish EPA, 2020).
- **Germany**: Germany is another leader in waste management, with one of the most successful recycling programs in the world. The country has a **dual system** for recycling, where households and businesses separate recyclables from waste. Germany's **Green Dot system** encourages manufacturers to take responsibility for the

packaging waste they generate, requiring them to pay for recycling programs. The country has also invested in waste-to-energy technology and strives for a closed-loop system where materials are reused instead of disposed of (Ritcher et al., 2019).

- **Japan**: Japan is known for its efficient waste management systems, particularly in urban areas like Tokyo. The country has implemented strict waste separation regulations, requiring citizens to separate their waste into categories such as burnable, non-burnable, and recyclable. This approach has led to high recycling rates and reduced waste going to landfills. Japan has also pioneered **waste-to-energy** technologies, turning waste into electricity through advanced incineration and other processes (Ueda et al., 2018).
- **San Francisco, USA**: San Francisco is one of the cities in the United States at the forefront of waste reduction and recycling. The city has adopted ambitious goals to achieve **zero waste** by 2020, and it has made significant progress by implementing comprehensive recycling and composting programs. In 2020, the city achieved a diversion rate of over 80%, meaning that more than 80% of waste is diverted from landfills through recycling, composting, or repurposing materials (SF Environment, 2020).
- **South Korea**: South Korea has successfully implemented a waste management program that encourages citizens to reduce waste and recycle. The **volume-based waste fee system** charges residents based on the amount of waste they dispose of, incentivizing recycling and waste reduction. The country has also invested heavily in **waste-to-energy** technologies and has one of the highest

recycling rates in the world (Park et al., 2020).

Conclusion: Moving Toward Sustainable Waste Management

The global waste crisis poses significant environmental challenges, but the concept of the circular economy offers a viable solution. By adopting waste reduction strategies, promoting recycling, and rethinking the design of products, countries and cities can reduce their environmental impact and conserve valuable resources. Leading examples from Sweden, Germany, Japan, San Francisco, and South Korea highlight the potential for innovative waste management practices to make a meaningful difference in the fight against pollution and environmental degradation.

The Role of Education and Awareness

Education and awareness are key drivers of environmental change. As the global environmental crisis intensifies, it is crucial that individuals, communities, businesses, and governments understand the issues and take collective action to address them. By fostering environmental consciousness and promoting sustainable behaviors, education can empower individuals to make informed decisions that benefit both the planet and society. Awareness campaigns and environmental education programs have the potential to create long-term, meaningful change by shaping attitudes, behaviors, and policies that support a sustainable future.

How Education and Awareness Campaigns Can Drive Change

Education and awareness campaigns have proven to be effective tools for initiating social change. By providing accessible information about the environment and the consequences of human actions, these campaigns can inspire individuals to adopt more sustainable lifestyles. Some ways in which education can drive environmental change include:

- **Raising Awareness**: Environmental education helps raise awareness about key issues such as climate change, pollution, resource depletion, and biodiversity loss. By disseminating information through schools, media, and community programs, individuals are better equipped to understand the urgency of addressing these challenges. Awareness campaigns can highlight the connections between personal choices (e.g., energy consumption, waste generation) and global environmental issues.
- **Shifting Behaviors**: Education fosters behavioral change by teaching individuals about the impact of their daily actions on the environment. Campaigns and programs that focus on recycling, energy conservation, water use reduction, and sustainable transportation can encourage people to adopt practices that minimize their ecological footprint. In some cases, small lifestyle changes can lead to significant cumulative benefits when practiced on a large scale.
- **Influencing Policy**: Education and awareness also play a critical role in shaping public opinion and influencing policy decisions. When individuals are educated about the environmental impacts of certain industries, products, or practices, they can advocate for stronger environmental regulations and policies. As public pressure grows, governments and businesses may be more likely to prioritize sustainability in their decisions. For example, campaigns advocating for renewable energy or plastic waste reduction have led to policy changes in various countries.
- **Building Environmental Literacy**: Environmental education provides individuals with the knowledge and skills necessary to engage in meaningful conversations

about sustainability. Environmental literacy enables people to critically evaluate the environmental implications of their decisions, both in their personal lives and in their communities. By fostering a deep understanding of ecological systems, resource management, and environmental justice, education prepares individuals to address environmental issues in an informed and responsible way.

The Impact of Environmental Education on Individuals and Communities

Environmental education has a profound impact on individuals and communities, fostering a sense of responsibility and empowering people to take action in their own lives. Some key outcomes of environmental education include:

- **Increased Environmental Consciousness**: Education helps individuals recognize the interconnectedness of human society and the natural world. People who are educated about environmental issues are more likely to understand the importance of protecting natural resources, conserving energy, reducing waste, and supporting sustainable practices. This awareness often leads to lifestyle changes, such as adopting plant-based diets, reducing plastic consumption, or choosing renewable energy options.
- **Community Engagement**: Environmental education can unite communities around common goals and inspire collective action. By working together on projects such as community gardens, local clean-up events, or recycling programs, individuals can strengthen their connections to one another and their shared environment. As

communities engage in these efforts, they also create social norms that promote sustainability and environmental stewardship.

- **Empowerment Through Knowledge**: Environmental education empowers individuals with the tools and knowledge they need to take positive action. By learning about climate change, resource management, and conservation, individuals feel more capable of making informed choices that benefit the environment. This sense of empowerment can lead to greater participation in environmental advocacy and decision-making processes.
- **Improved Health and Well-being**: Environmental education often encourages practices that promote health and well-being, such as reducing pollution, creating green spaces, and eating sustainably sourced food. Communities that prioritize environmental education tend to see improvements in public health, as these practices often reduce exposure to toxins, improve air and water quality, and encourage physical activity in natural settings.

The Rise of Grassroots Movements Advocating for Environmental Responsibility

Grassroots movements have played a pivotal role in advocating for environmental responsibility, often driving significant social and political change. These movements are typically led by individuals, community organizations, or local groups that work at the ground level to raise awareness, mobilize action, and demand policy changes. Some key aspects of grassroots movements include:

- **Grassroots Activism**: Many of today's most successful environmental campaigns have emerged from grassroots activism. Activists, often from marginalized communities, organize protests, petitions, and awareness campaigns to demand government action on environmental issues. The rise of global movements like **Fridays for Future**, led by young climate activist **Greta Thunberg**, exemplifies how grassroots efforts can influence public discourse and policy at the highest levels.

- **Local Initiatives**: Grassroots movements often focus on local issues, such as waste reduction, sustainable agriculture, or clean energy projects. By addressing environmental concerns at the local level, these movements encourage individuals to take ownership of their surroundings and foster a sense of community responsibility. Local initiatives can also inspire larger-scale changes by demonstrating the viability of sustainable practices in everyday life.
- **Youth Activism**: The involvement of youth in environmental movements has surged in recent years, with young people advocating for urgent action on climate change, pollution, and other environmental issues. Youth-led organizations and campaigns, such as **Extinction Rebellion** and **Sunrise Movement**, have gained significant traction, urging governments to prioritize sustainability and climate justice. These movements are shaping the future of environmental activism by giving young people a platform to voice their concerns and demand change.
- **Global Solidarity**: Grassroots movements often emphasize the importance of global solidarity in addressing environmental challenges. They call for collective action across borders to tackle issues like climate change, biodiversity loss, and pollution. Movements like **350.org**, which campaigns for reducing carbon dioxide

levels to a safe level of 350 parts per million, demonstrate how grassroots efforts can unite people around a shared vision of a sustainable future.

Conclusion: Education as a Catalyst for Change

Education and awareness campaigns are powerful tools in addressing the environmental crisis. By raising awareness, shifting behaviors, and empowering individuals and communities, environmental education can drive significant change. Grassroots movements, fueled by a sense of responsibility and urgency, are helping to reshape the global conversation about sustainability and environmental justice. Together, education, awareness, and activism offer hope for a more sustainable future.

Corporate Responsibility and Green Business Practices

Corporations play a significant role in both driving environmental degradation and offering solutions for sustainability. Over the past few decades, businesses have become increasingly aware of their environmental impact and the need for responsible practices. As major consumers of resources and producers of waste, companies are in a unique position to lead the charge in reducing their ecological footprints. The growing focus on **corporate responsibility** has led many businesses to adopt **green business practices** aimed at reducing their environmental impact while also benefiting from the long-term advantages of sustainability.

The Role of Corporations in Environmental Degradation

Corporations have historically been major contributors to environmental degradation, particularly through their industrial activities, resource extraction, and waste production. Some of the key ways in which businesses have contributed to environmental harm include:

- **Resource Extraction**: Many industries, such as mining,

forestry, and oil and gas extraction, contribute to environmental degradation through unsustainable resource use. Deforestation, habitat destruction, and water pollution are some of the direct consequences of resource extraction activities, often driven by the demand for raw materials used in manufacturing and production.

- **Greenhouse Gas Emissions**: Large corporations in energy-intensive sectors like manufacturing, transportation, and agriculture are significant contributors to **greenhouse gas emissions**. These emissions contribute to global warming, leading to changes in climate patterns, rising sea levels, and increased frequency of extreme weather events.
- **Waste Generation**: The global production and consumption model, driven by corporations, has led to an enormous amount of waste. Many companies produce single-use plastic, packaging, and products with short lifespans that end up in landfills or polluting oceans. The disposal of waste, particularly non-biodegradable materials, is a significant environmental issue linked to corporate practices.
- **Pollution**: Corporations are responsible for a large portion of air, water, and soil pollution. Industrial activities release toxic chemicals, heavy metals, and particulate matter into the environment, damaging ecosystems and human health. These pollutants often end up in rivers, lakes, oceans, and the atmosphere, causing long-term harm to the planet.

While corporations have historically been major contributors to environmental degradation, they also have the potential to significantly reduce their impact by adopting sustainable practices and taking on greater corporate responsibility.

Examples of Companies That Have Adopted Sustainable Business Practices

As awareness of environmental issues grows, many companies are increasingly adopting green business practices, recognizing that sustainability not only benefits the planet but also enhances their profitability and reputation. Some companies have made substantial strides in reducing their environmental impact through eco-friendly initiatives, sustainable supply chains, and responsible business practices.

- **Patagonia**: A well-known leader in environmental responsibility, **Patagonia** has consistently focused on sustainability as a core aspect of its business model. The outdoor clothing company promotes the use of recycled materials, encourages customers to buy used clothing through its **Worn Wear** program, and donates a percentage of profits to environmental causes. Additionally, Patagonia has taken a stand on issues like climate change and actively advocates for environmental protection through campaigns and political lobbying.
- **Unilever: Unilever**, a global consumer goods company, has committed to achieving net-zero emissions by 2039. The company has integrated sustainability into its operations, with a focus on reducing water usage, improving waste management, and sourcing sustainable raw materials. Unilever's commitment to sustainability is also reflected in its product lines, which include eco-friendly products like biodegradable detergents and plant-based food products. Furthermore, Unilever has worked to ensure that its supply chain adheres to ethical labor and environmental

standards, promoting a circular economy and reducing waste.

- **Tesla**: **Tesla** has made a significant impact on the automotive industry by producing electric vehicles that help reduce reliance on fossil fuels and decrease carbon emissions. The company has also developed solar energy products and energy storage solutions, contributing to the transition to renewable energy. Tesla's mission is to accelerate the world's shift toward sustainable energy, and its products are designed to reduce environmental harm, from electric cars to solar panels. Tesla's green business practices align with its commitment to combat climate change by promoting renewable energy and energy efficiency.
- **IKEA: IKEA** has made considerable efforts to adopt sustainable practices across its operations, from sourcing renewable and recycled materials to reducing its carbon footprint. The company has committed to becoming climate positive by 2030, meaning that it will reduce more greenhouse gases than its value chain emits. IKEA has also implemented a circular business model, offering services such as furniture repair and refurbishment, as well as encouraging customers to recycle their products. Additionally, the company uses renewable energy in its operations and aims to make all of its products from renewable or recycled materials by 2030.
- **Google**: As a technology company, **Google** has taken significant steps toward sustainability, including achieving **carbon neutrality** in its operations. Google has committed to running its data centers on renewable energy and has made substantial investments in clean energy projects. The

company also works to reduce waste by promoting the circular economy through the recycling of old hardware and the design of energy-efficient devices. Google is also committed to reducing its environmental impact through its supply chain, partnering with suppliers that share its commitment to sustainability.

- **Microsoft**: Microsoft is another major corporation that has committed to reducing its environmental impact. The company has pledged to become carbon negative by 2030, which means it will remove more carbon from the atmosphere than it emits. Microsoft is investing in renewable energy, carbon capture technology, and sustainable practices in its supply chain. Additionally, the company has committed to reducing water usage and achieving zero waste in its operations. Microsoft's focus on sustainability is also evident in its products and services, such as cloud computing solutions that help other organizations reduce their carbon footprints.

Conclusion: The Growing Importance of Corporate Responsibility

Corporations play an undeniable role in environmental degradation, but they also have the potential to lead the way toward sustainability. By adopting green business practices, companies can reduce their environmental impact while reaping the benefits of improved efficiency, customer loyalty, and brand reputation. The examples of companies like Patagonia, Unilever, Tesla, IKEA, Google, and Microsoft show that it is possible for businesses to integrate sustainability into their operations and contribute to the global effort to combat environmental degradation. As environmental awareness continues to grow, the pressure on corporations to act responsibly will increase, and businesses that prioritize sustainability will be better positioned to thrive in a future that values environmental stewardship.

The Role of Individuals in Addressing the Crisis

While corporations and governments play critical roles in tackling the global environmental crisis, individuals also have a significant part to play. The choices that people make in their daily lives, from the products they purchase to the energy they consume, can either contribute to environmental degradation or help drive positive change. Individuals have the power to make impactful decisions that collectively contribute to the sustainability of the planet. When people adopt environmentally conscious behaviors, raise awareness, and engage in activism, they help create a culture of sustainability and push for systemic change.

How Individuals Can Contribute to Solving the Environmental Crisis

Individuals can contribute to solving the environmental crisis in a variety of ways, from making personal lifestyle changes to getting involved in community and global initiatives. These actions, while seemingly small in isolation, add up when practiced by millions of people around the world. Some of the key actions that individuals can take include:

- **Reducing Waste**: One of the most effective ways individuals can reduce their environmental impact is by reducing waste. This can be achieved through actions like recycling, composting organic matter, reducing single-use plastic, and choosing products with minimal packaging. Practicing **zero-waste** living—buying in bulk, repurposing items, and avoiding disposable products—helps conserve resources and reduces pollution. Additionally, reducing food waste by planning meals and using leftovers can help reduce the amount of waste sent to landfills, thereby decreasing methane emissions.

- **Conserving Energy**: Energy conservation is another crucial area where individuals can make a difference. By reducing energy consumption in the home, people can decrease their carbon footprints and reduce the demand for fossil fuels. Simple actions like turning off lights when not in use, using energy-efficient appliances, and insulating homes to minimize heating and cooling needs all contribute to energy savings. Switching to renewable energy sources, such as solar power, can further reduce reliance on fossil fuels. Many individuals also make the shift to electric vehicles (EVs) as part of their commitment to reducing carbon emissions.
- **Supporting Eco-friendly Products**: Purchasing products that are sustainably made, ethically sourced, and have minimal environmental impact is another way individuals can contribute to the environmental crisis. Supporting businesses that prioritize **sustainable sourcing, eco-friendly manufacturing processes**, and **fair labor practices** encourages more companies to adopt environmentally responsible practices. Choosing organic food, avoiding products with harmful chemicals, and buying from companies that commit to reducing their environmental impact are all ways individuals can align their purchasing habits with their values.
- **Adopting Sustainable Transportation**: Individuals can also reduce their environmental impact by choosing more sustainable transportation options. This includes walking, cycling, using public transit, or carpooling instead of driving alone. For those who need to drive, opting for an electric or hybrid vehicle can reduce greenhouse gas

emissions. In addition, individuals can advocate for better transportation infrastructure in their communities, such as bike lanes and public transit options, which encourage sustainable mobility.

The Importance of Activism and Political Engagement

While individual lifestyle changes are important, activism and political engagement are essential for addressing the broader systemic issues driving environmental degradation. Individuals who advocate for environmental protection can help influence government policies, corporate practices, and public attitudes toward sustainability. By becoming active participants in the environmental movement, people can demand stronger actions on climate change, pollution, deforestation, and other critical issues. Some of the ways individuals can engage in activism and political engagement include:

- **Advocating for Policy Change**: Individuals can use their voices to demand stronger environmental regulations and policies from their governments. This may include calling for stricter emissions standards, supporting the transition to renewable energy, advocating for the protection of public lands, and promoting sustainable agriculture practices. Individuals can write letters to elected officials, sign petitions, participate in protests, or vote for political candidates who prioritize environmental sustainability.
- **Joining or Supporting Environmental Movements**: Many grassroots environmental movements have emerged around the world to address urgent issues like climate change and environmental justice. By joining these movements or supporting their causes, individuals can amplify their impact and help mobilize communities to

take collective action. Movements such as **Fridays for Future**, **Extinction Rebellion**, and **350.org** have gained widespread support and continue to call for immediate action to protect the planet. These movements rely on the engagement of passionate individuals who are committed to raising awareness and pushing for change.

- **Engaging in Community-based Projects**: Local activism can have a powerful impact on the environment. Individuals can engage in community-based environmental projects, such as organizing clean-up events, planting trees, creating community gardens, and promoting local sustainability initiatives. These efforts not only improve the environment but also foster a sense of collective responsibility and solidarity.
- **Supporting Environmental Organizations**: Individuals can contribute to the fight against environmental degradation by supporting organizations that work to protect the planet. This can include donating time, resources, or money to NGOs that focus on conservation, climate change, renewable energy, and other environmental issues. Organizations like **Greenpeace**, **World Wildlife Fund (WWF)**, and **The Sierra Club** rely on public support to fund their campaigns and advocacy work. By backing these groups, individuals can help amplify their efforts to create meaningful change.
- **Raising Awareness on social media**: In today's digital age, social media can be a powerful tool for raising awareness and driving action. Individuals can use platforms like Instagram, Twitter, and Facebook to share information about environmental issues, highlight sustainable products and practices, and engage with others who are passionate

about the environment. Social media campaigns can rapidly spread information and rally people to take action, whether it's by participating in global climate strikes, donating to environmental causes, or simply spreading the message about sustainability.

Conclusion: Collective Action for a Sustainable Future

The environmental crisis may feel overwhelming at times, but individual actions—whether through lifestyle changes, activism, or political engagement—have the power to make a significant difference. By reducing waste, conserving energy, supporting eco-friendly businesses, and engaging in environmental advocacy, individuals can help drive the change needed to protect the planet for future generations. Collective action, fueled by a shared commitment to sustainability, is key to solving the global environmental crisis.

Hope for the Future: A Call to Action

While the environmental crisis presents some of the greatest challenges humanity has ever faced, there are reasons for hope. Around the world, individuals, organizations, and governments are stepping up to tackle environmental degradation, and successful efforts to reduce environmental impact are already being seen. These efforts demonstrate that a sustainable future is possible, provided that we take urgent action. There is a growing global movement toward sustainability that is driven by innovation, collaboration, and a shared vision of a healthier planet for future generations.

Highlight Successful Efforts in Reducing Environmental Impact

Several examples from around the world illustrate the positive changes that are possible when concerted efforts are made to address environmental issues. These success stories provide inspiration and hope that meaningful progress can be achieved.

- **The Transition to Renewable Energy**: Several countries and regions have made significant strides in shifting from fossil fuels to renewable energy sources like wind, solar, and hydropower. **Iceland**, for example, generates nearly 100% of its electricity from renewable sources, primarily geothermal and hydroelectric power. Similarly, **Denmark** is a global leader in wind energy, with wind turbines supplying around 50% of the country's electricity needs. The rapid growth of renewable energy technologies is making it possible to reduce dependence on fossil fuels and cut carbon emissions, contributing to the fight against climate change.
- **Reforestation and Land Restoration**: Reforestation and land restoration projects are proving to be highly effective in combating deforestation, restoring ecosystems, and capturing carbon from the atmosphere. For example, **China** has undertaken massive reforestation projects, planting billions of trees over the past few decades. These efforts are helping to restore degraded land, improve air quality, and protect biodiversity. Other countries, such as **Costa Rica**, have also implemented successful reforestation programs that have not only reduced deforestation but also contributed to the recovery of endangered species and the preservation of natural habitats.
- **Circular Economy Initiatives**: Some cities and companies are embracing the concept of a circular economy, where waste is minimized, and resources are reused and recycled. **Amsterdam** has set ambitious goals to become a fully circular economy by 2050, encouraging businesses to design products with longer lifespans and greater recyclability. This model not only reduces waste but also

conserves resources and reduces the carbon footprint associated with the production of new goods.

- **Sustainable Agriculture and Food Systems**: Sustainable agricultural practices are making a difference in reducing the environmental impact of food production. Practices such as regenerative farming, organic agriculture, and permaculture are gaining traction and are proving to be effective in enhancing soil health, increasing biodiversity, and reducing the need for chemical fertilizers and pesticides. **The Savory Institute**, for instance, has been promoting **holistic management** practices that focus on restoring ecosystems through better land management. These practices help mitigate climate change by sequestering carbon in the soil and improving food security for communities.
- **Global Climate Agreements and Policy Change**: International agreements such as the **Paris Agreement** have been a major milestone in global efforts to combat climate change. Countries around the world have committed to limiting global warming to well below 2°C, with the ambition of striving for 1.5°C. While challenges remain, the agreement represents a collective acknowledgment of the need for urgent action and cooperation across borders. Local governments are also taking action, such as **California's** commitment to achieving carbon neutrality by 2045, which is influencing other regions and countries to follow suit.

A Message of Hope for a Sustainable Future

Despite the enormity of the environmental challenges we face, there is a strong message of hope for the future. The progress made so far shows that when people and governments prioritize sustainability, positive change is possible. The increasing use of renewable energy, the success of reforestation efforts, and the adoption of circular economy principles all provide evidence that we can create a sustainable future for ourselves and future generations.

However, it is crucial to recognize that the scale of the crisis requires continued and accelerated action. While these successes are promising, they must be expanded upon to meet the urgent challenges posed by climate change, biodiversity loss, and resource depletion. The window of opportunity to mitigate the worst effects of the environmental crisis is closing, but with swift and sustained action, a sustainable future is within reach.

The Importance of Collective Effort in Mitigating the Crisis

Addressing the environmental crisis requires a collective effort—no single person, organization, or country can solve the problem alone. It will take the collaboration of governments, businesses, civil society, and individuals to implement the necessary changes at the scale required to make a lasting impact. Individuals have a vital role to play by adopting sustainable lifestyles, advocating for policy change, and supporting environmental causes. Governments must lead by enacting and enforcing policies that promote sustainability, invest in green technologies, and protect natural resources. Businesses must take responsibility for their environmental footprint and adopt sustainable practices that contribute to a circular economy. Non-governmental organizations, communities, and grassroots movements are crucial in holding governments and corporations accountable and raising awareness about the urgent need for action.

The success of collective efforts can be seen in global movements like **Fridays for Future, the Extinction Rebellion**, and the growing demand for climate action from youth activists around the world. These movements are empowering individuals to take a stand for the environment, pressuring governments to act, and inspiring others to join the cause.

A Call to Action

The environmental crisis is a defining challenge of our time, but it is also an opportunity to build a more sustainable, just, and resilient world. From the successes in renewable energy adoption to the global movements for climate action, there is hope that humanity can overcome the environmental crisis and create a brighter future.

Now is the time for action. Every individual, community, business, and government must contribute to the effort to mitigate the environmental crisis. By making conscious choices, advocating for change, and working together, we can reduce our impact on the planet and ensure a sustainable future for generations to come. The crisis may be vast, but our collective will to act is even more powerful.

Conclusion

The environmental crisis, driven by climate change, pollution, resource depletion, and loss of biodiversity, is one of the most significant challenges facing humanity today. Throughout this paper, we have explored the historical roots of environmental degradation, the impacts of climate change on ecosystems and human societies, and the urgent need for action to address these critical issues. We have examined the role of human population growth, political and economic drivers, as well as the disproportionate impacts on poorer nations, and highlighted the importance of global cooperation and individual responsibility in mitigating the crisis.

The environmental degradation we face today is a result of centuries of industrialization, unsustainable resource use, and political and economic systems that prioritize short-term gains over long-term sustainability. The consequences of inaction are already being felt, from rising global temperatures to extreme weather events and the loss of species. However, there is still hope, as demonstrated by successful efforts in renewable energy adoption, sustainable agriculture, reforestation, and policy change. These examples show that progress is possible when we come together to confront the crisis.

The Urgency of Addressing the Environmental Crisis

The urgency of addressing the environmental crisis cannot be overstated. The scientific consensus on the causes and impacts of climate change, biodiversity loss, and resource depletion is clear: we are rapidly approaching critical tipping points. The window for taking effective action to avoid the most catastrophic consequences is closing, and immediate, comprehensive steps must be taken to mitigate the crisis. From transitioning to renewable energy to protecting ecosystems and reducing waste, the actions we take today will shape the future of our planet.

The time to act is now. Every fraction of a degree of temperature rise matters, and the longer we delay, the harder it will be to avoid irreversible damage to the planet's ecosystems and the health and livelihoods of people worldwide. Governments, businesses, and individuals all have a responsibility to contribute to the solution, and the decisions made in the coming years will determine the trajectory of our collective future.

A Call to Action for Global Cooperation and Individual Responsibility

Addressing the environmental crisis requires a collective effort. Governments must enact and enforce policies that prioritize sustainability, protect natural resources, and drive the transition to a green economy. Businesses must take responsibility for their environmental footprints and adopt practices that reduce waste, conserve resources, and support circular economies. Non-governmental organizations, grassroots movements, and local communities must continue to advocate for environmental justice and hold leaders accountable for their actions.

However, the role of individuals is equally important. Each of us has the power to make sustainable choices in our daily lives, whether by reducing waste, conserving energy, or supporting eco-friendly products and businesses. Through activism, education, and political engagement, individuals can contribute to the growing movement for environmental protection, pressuring governments and corporations to take stronger action. The collective will of millions of individuals, each making small changes, can add up to a significant transformation.

In conclusion, the environmental crisis is not an insurmountable challenge but a call to action for all of us. It is an opportunity to work together, across borders and sectors, to create a more sustainable and resilient world. The future of the planet depends on our ability to act now, with urgency, unity, and responsibility. Together, we can mitigate the damage and build a future where people and nature thrive in harmony.

Chapter 1: Climate Change: Causes and consequences

Introduction to Climate Change Climate change refers to significant, long-term changes in temperature, weather patterns, and other aspects of the Earth's climate. While the Earth's climate has naturally fluctuated throughout history, recent warming trends are largely attributed to human activities. The driving force behind contemporary climate change is the increased concentration of greenhouse gases (GHGs) in the atmosphere. These gases trap heat, leading to global warming and subsequent shifts in weather patterns. As the world's population grows and industrialization continues, the burning of fossil fuels, deforestation, and agricultural practices have accelerated the release of carbon dioxide (CO2) and other GHGs. In addition to rising global temperatures, the effects of climate change include melting ice caps, rising sea levels, and an increase in extreme weather events such as floods, hurricanes, and heatwaves. Understanding the causes and consequences of climate change is crucial in developing effective mitigation strategies. The scientific community has widely acknowledged that human activities are the primary contributors to climate change, and urgent action is necessary to reduce emissions and limit its impacts (Intergovernmental Panel on Climate Change. This chapter aims to explore the causes of climate change, its consequences, and the steps that need to be taken to address this pressing global challenge.

The Science Behind Climate Change The science behind climate change revolves around the Earth's climate system, which is governed by the exchange of energy between the Earth, the atmosphere, and the sun. The Earth's climate has naturally fluctuated over millions of years due to factors like volcanic activity, solar radiation, and changes in the Earth's orbit. However, since the industrial revolution, human activities have significantly altered the composition of the atmosphere, particularly by increasing the concentration of greenhouse gases (GHGs) such as carbon dioxide (CO_2), methane (CH_4), and nitrous oxide (N_2O). These gases trap heat in the atmosphere, leading to the greenhouse effect, which is the primary cause of global warming (National Aeronautics and Space Administration [NASA], 2020). Scientific studies indicate that the Earth's surface temperature has risen by approximately 1.2°C since the late 19th century, with the rate of warming accelerating in recent decades. The increase in GHGs is largely driven by the burning of fossil fuels for energy, deforestation, and industrial processes. The consequences of this warming are far-reaching, including rising sea levels, more frequent and intense heatwaves, and shifting weather patterns. Understanding the science behind climate change is essential for developing mitigation strategies, such as reducing emissions and transitioning to renewable energy sources. Climate models predict that without significant intervention, global temperatures will continue to rise, exacerbating the effects of climate change (Hansen et al., 2008).

The Role of Greenhouse Gases Greenhouse gases (GHGs) are responsible for trapping heat in the Earth's atmosphere, a process known as the greenhouse effect. The primary GHGs contributing to climate change are carbon dioxide (CO_2), methane (CH_4), nitrous oxide (N_2O), and fluorinated gases. CO_2 is the most abundant GHG released by human activities, primarily from the combustion of fossil fuels such as coal, oil, and natural gas for energy production and transportation. Methane, though present in smaller quantities, is a much more potent GHG, with a global warming potential approximately 25 times greater than CO_2 over a 100-year period (Intergovernmental Panel on Climate Change. It is released during the production and transport of coal, oil, and natural gas, as well as from agricultural practices, particularly livestock farming. Nitrous oxide, mainly emitted from agricultural and industrial activities, is another significant GHG that contributes to warming. Fluorinated gases, including hydrofluorocarbons (HFCs), are synthetic gases used in refrigeration and air conditioning. These gases have a much higher global warming potential, though they are present in much smaller quantities. The buildup of these gases in the atmosphere is accelerating due to industrialization, leading to the enhanced greenhouse effect, which is causing global temperatures to rise. Reducing emissions of GHGs is critical to mitigating climate change and limiting global warming to safe levels.

Human Activities and Greenhouse Gas Emissions

Human activities are the primary drivers of the current climate change crisis. The significant increase in greenhouse gas (GHG) emissions since the Industrial Revolution is a direct result of human activities such as burning fossil fuels, deforestation, and industrial processes. Fossil fuels, including coal, oil, and natural gas, have been central to powering industries, transportation, and energy production, but they release large amounts of CO2 and other GHGs when burned. In addition, deforestation and land-use changes contribute significantly to climate change by reducing the number of trees that can absorb CO2 from the atmosphere. The loss of forests exacerbates the problem by releasing the stored carbon back into the atmosphere, further accelerating global warming. Agriculture, particularly livestock farming, also plays a major role by producing methane, a potent GHG. Fertilizers used in farming release nitrous oxide, another GHG with a high global warming potential. Moreover, industrial processes, including cement production and the manufacture of chemicals and metals, release additional CO2 and other GHGs. These human activities have created an imbalance in the Earth's natural carbon cycle, which has accelerated climate change and created a warming trend that is now felt worldwide.

Reducing emissions from these human activities is crucial to mitigating climate change and limiting its impacts.

Burning Fossil Fuels and its Impact

The burning of fossil fuels is the primary contributor to the rising levels of greenhouse gases (GHGs) in the atmosphere, and consequently, to the global warming that defines climate change. Fossil fuels, which include coal, oil, and natural gas, are burned to generate energy for electricity production, industrial processes, transportation, and residential heating. When burned, fossil fuels release CO2, a major greenhouse gas that traps heat in the atmosphere and contributes to global warming. Coal is the most carbon-intensive fossil fuel, followed by oil and natural gas, although all three contribute significantly to climate change. The global demand for energy has continued to rise, and fossil fuels remain the dominant source of energy worldwide, despite the increasing push for renewable energy solutions. This reliance on fossil fuels has led to an unprecedented rise in CO2 concentrations in the atmosphere, with current levels far exceeding those of previous geological eras. The effects of this sustained increase in CO2 are already evident, with rising global temperatures, melting ice caps, and more frequent extreme weather events. Efforts to mitigate the impact of burning fossil fuels include transitioning to cleaner, renewable energy sources, implementing energy efficiency measures, and investing in carbon capture and storage technologies. However, substantial reductions in fossil fuel consumption are essential to achieving the targets set by international climate agreements such as the Paris Agreement.

Deforestation and Land Use Changes

Deforestation and land use changes are among the most significant human-induced drivers of climate change. Forests act as carbon sinks, meaning they absorb more carbon dioxide (CO2) from the atmosphere than they release. When forests are cleared for agriculture, urbanization, or logging, this carbon sequestration process is disrupted, and the carbon stored in the trees is released back into the atmosphere, contributing to global warming (Bastin et al., 2019). In addition to CO2 emissions from deforestation, the destruction of forests also diminishes the Earth's capacity to absorb future emissions, creating a feedback loop that exacerbates climate change. Tropical rainforests, such as the Amazon, are particularly crucial in regulating the global carbon cycle and maintaining biodiversity. However, they are rapidly disappearing due to human activities. Agricultural expansion, particularly for cattle ranching and the cultivation of crops like soy and palm oil, is a major cause of deforestation. Similarly, urban sprawl and infrastructure development also lead to land-use changes that further diminish the carbon storage capacity of forests. In addition to the direct impact on climate change, deforestation contributes to biodiversity loss, disrupts local water cycles, and affects the livelihoods of Indigenous communities who depend on forests for their survival. Combating deforestation requires a combination of policies, including stricter regulations on land use, reforestation efforts, and support for sustainable land management practices.

Global Warming and Temperature Rise

Global warming refers to the long-term increase in the Earth's average surface temperature due to human-induced factors, particularly the accumulation of greenhouse gases (GHGs) in the atmosphere. Since the late 19th century, the Earth's average temperature has risen by approximately 1.2°C, with the past few decades experiencing accelerated warming. The primary driver of global warming is the enhanced greenhouse effect, which occurs when GHGs such as carbon dioxide (CO2), methane (CH4), and nitrous oxide (N2O) trap heat in the atmosphere. This trapped heat causes the Earth's surface to warm, resulting in higher average global temperatures. The consequences of this warming are already being observed, with more frequent and intense heatwaves, the melting of polar ice caps, and the loss of glaciers. The impacts of global warming are far-reaching, affecting ecosystems, agriculture, water resources, and human health. A rise of even a few degrees in global temperature can lead to significant changes in climate patterns, including more intense storms, floods, droughts, and shifts in ecosystems. The scientific consensus is clear: continued global warming will lead to even more severe consequences if left unchecked. Limiting temperature rise to well below 2°C, as outlined in the Paris Agreement, will require immediate and substantial reductions in GHG emissions across all sectors.

Impact on Polar Ice and Glaciers

The impact of climate change on polar ice and glaciers is one of the most visible and alarming consequences of global warming. Rising global temperatures are causing the melting of ice sheets and glaciers, particularly in the Arctic and Antarctic regions. Polar ice plays a critical role in regulating the Earth's climate by reflecting sunlight back into space, a process known as the albedo effect. As the ice melts, darker ocean or land surfaces are exposed, absorbing more heat and further accelerating the warming process. In addition to contributing to rising sea levels, the melting of polar ice and glaciers disrupts delicate ecosystems that depend on cold environments. For example, polar bears and other Arctic species are at risk as their habitats are diminished by the shrinking ice. The loss of glaciers in mountain regions also threatens freshwater sources for millions of people who rely on glacial meltwater for drinking, agriculture, and hydropower. Recent studies show that the rate of ice loss from Antarctica has more than tripled since the 1970s, and the Arctic is warming at more than twice the global average rate. If these trends continue, the consequences could be catastrophic for both ecosystems and human populations. To mitigate these impacts, urgent action to reduce global emissions and slow the rate of warming is essential.

Sea-Level Rise and Coastal Flooding

Sea-level rise is one of the most significant consequences of climate change, with profound implications for coastal communities and ecosystems worldwide. The two primary contributors to sea-level rise are the thermal expansion of seawater due to global warming and the melting of polar ice and glaciers. As the Earth's temperature increases, the ocean absorbs excess heat, causing water to expand and increase its volume. At the same time, the melting of ice sheets in Greenland and Antarctica, as well as mountain glaciers, contributes additional water to the oceans, further elevating sea levels. As a result, many low-lying coastal areas are already experiencing more frequent and severe flooding, as well as increased salinity in freshwater resources. Cities like Miami, New York, and Jakarta are facing the threat of regular inundation, which could result in the displacement of millions of people in the coming decades. Additionally, rising sea levels pose a significant threat to vital ecosystems, such as mangroves, wetlands, and coral reefs, which provide critical services including carbon sequestration, coastal protection, and biodiversity. Adaptation strategies such as building sea walls, restoring wetlands, and relocating vulnerable communities are crucial to managing the impacts of sea-level rise, but ultimately, reducing global emissions remains the most effective long-term solution.

Extreme Weather Events

Extreme weather events, including hurricanes, heatwaves, floods, and droughts, are becoming more frequent and intense as a result of climate change. The warming of the Earth's atmosphere alters the patterns of weather systems, leading to an increase in the occurrence of these extreme events. For example, higher temperatures contribute to more intense heatwaves, which can have devastating impacts on human health, agriculture, and energy systems. Similarly, warmer air holds more moisture, which can lead to heavier rainfall and flooding, particularly in regions that are already vulnerable to such events. The frequency of intense hurricanes and storms has also increased, with warmer sea surface temperatures providing more energy for these systems to develop. The 2020 Atlantic hurricane season, for example, broke records for the number of named storms, and this trend is expected to continue as global temperatures rise. In contrast, some regions are experiencing more severe droughts due to changing precipitation patterns and higher evaporation rates, which threaten water supplies, food security, and agriculture. Extreme weather events are not only destructive but also disproportionately affect vulnerable populations, particularly those in developing countries, who lack the resources to prepare for and recover from such disasters. To reduce the impact of extreme weather, efforts to limit global warming and increase resilience through improved infrastructure and disaster response are essential.

Ecosystem Disruption

Climate change is causing widespread disruption to ecosystems around the world, affecting biodiversity, species distributions, and the functioning of natural systems. Rising temperatures, changes in precipitation patterns, and the increasing frequency of extreme weather events are forcing many species to adapt, migrate, or face extinction. For example, warming oceans are causing coral reefs to bleach and die, threatening the biodiversity that depends on these ecosystems. Similarly, warmer temperatures are shifting the ranges of many species, forcing them to move toward cooler regions, often at the expense of local populations. Additionally, ecosystems that are crucial for carbon sequestration, such as forests and wetlands, are being degraded by climate change, reducing their ability to absorb CO2 from the atmosphere. The loss of these ecosystems exacerbates climate change, creating a feedback loop that accelerates the problem. The disruption of ecosystems also has serious consequences for human societies, particularly in terms of food security, water availability, and disaster resilience. For example, the loss of pollinators, such as bees, due to changing weather patterns and habitat destruction, threatens global food production. Protecting ecosystems and biodiversity is crucial for both mitigating climate change and ensuring that natural systems continue to provide essential services for humanity.

Agriculture and Food Security

Agriculture is highly vulnerable to climate change, and shifts in weather patterns, rising temperatures, and extreme weather events have direct consequences for food security. Climate change affects crop yields, soil fertility, and water availability, all of which are essential for agricultural productivity. Changes in precipitation patterns can lead to droughts or floods, reducing the availability of water for irrigation and stressing crops. Warmer temperatures may also affect the growing season of crops, forcing farmers to adapt by altering planting schedules or selecting different crop varieties. However, these adaptations are not always feasible or effective, particularly in regions that are already struggling with poor infrastructure and limited resources. Extreme weather events, such as heatwaves, hurricanes, and floods, further exacerbate the challenges faced by farmers. For instance, heatwaves can cause crop failures by reducing the ability of plants to photosynthesize and grow, while floods can damage crops and infrastructure. Additionally, rising temperatures can lead to the spread of pests and diseases that threaten crop production. The resulting reductions in food supply, along with higher prices, could lead to increased food insecurity, particularly in developing countries where agriculture is a primary livelihood. Mitigating the impact of climate change on agriculture requires adopting sustainable farming practices, improving water management, and developing resilient crop varieties, as well as reducing emissions from the agriculture sector itself.

Health Impacts of Climate Change

The health impacts of climate change are far-reaching and pose significant risks to human well-being. Rising temperatures, changes in air quality, and extreme weather events directly affect both physical and mental health. Heatwaves, for example, can lead to heat stress, dehydration, and heatstroke, particularly among vulnerable populations such as the elderly, children, and those with pre-existing health conditions. In addition, climate change contributes to the spread of infectious diseases, as shifting temperature and rainfall patterns alter the habitats of disease-carrying vectors such as mosquitoes and ticks. Diseases like malaria, dengue fever, and Lyme disease are becoming more widespread in regions where they were previously uncommon. Air pollution, which worsens with higher temperatures, exacerbates respiratory conditions such as asthma and bronchitis, especially in urban areas. Floods and storms can also lead to injury, death, and the displacement of communities, further exacerbating health disparities. Mental health is also impacted by climate change, as people face stress and anxiety related to extreme weather events, loss of livelihoods, and displacement. Addressing the health impacts of climate change requires improving public health systems, increasing access to healthcare, and enhancing community resilience to climate-related risks.

The Role of Renewable Energy

Renewable energy plays a central role in combating climate change by providing an alternative to fossil fuels, which are the primary contributors to greenhouse gas (GHG) emissions. Sources of renewable energy include solar, wind, geothermal, and hydropower, all of which produce little or no direct emissions. Transitioning to renewable energy is essential to reducing global carbon emissions and mitigating climate change. Solar and wind energy, in particular, have seen significant technological advancements in recent years, making them increasingly cost-competitive with fossil fuels. The widespread adoption of these clean energy sources can reduce the reliance on coal, oil, and natural gas for electricity generation, transportation, and heating. Moreover, renewable energy offers a sustainable solution to energy poverty, providing electricity to remote and underserved regions without the need for expensive infrastructure or dependence on imported fuels. In addition to reducing emissions, renewable energy technologies help reduce air pollution, improve public health, and create new economic opportunities. However, challenges remain in scaling up renewable energy infrastructure, ensuring grid stability, and addressing intermittency issues, particularly for solar and wind power. Nevertheless, with the right policy frameworks and investments in research and development, renewable energy can play a pivotal role in a low-carbon future and contribute significantly to achieving the global climate targets set by agreements like the Paris Agreement.

International Climate Agreements and Policies

International climate agreements and policies are crucial to addressing the global challenge of climate change. The most significant of these agreements is the Paris Agreement, which was adopted by 196 countries in 2015 under the United Nations Framework Convention on Climate Change (UNFCCC). The Paris Agreement aims to limit global warming to well below 2°C above pre-industrial levels, with an aspiration to limit the rise to 1.5°C. Countries that are parties to the agreement have pledged to reduce their greenhouse gas (GHG) emissions and implement national climate action plans known as nationally determined contributions (NDCs). These commitments are reviewed and updated every five years, with the goal of progressively increasing ambition. The Paris Agreement also includes provisions for financial support to developing countries to help them mitigate and adapt to climate change. Other international agreements, such as the Kyoto Protocol (1997), have laid the groundwork for climate action by establishing binding emission reduction targets for industrialized countries. However, despite these agreements, progress has been uneven, and the world is currently not on track to meet the goals set by the Paris Agreement. Stronger policies, greater international cooperation, and more ambitious commitments are needed to avoid the worst impacts of climate change and ensure a sustainable future for all.

The Role of Individuals in Combating Climate Change

While governments and industries play a significant role in addressing climate change, individuals also have a crucial part to play in mitigating its impacts. The choices made by individuals, particularly regarding energy consumption, transportation, and waste management, can collectively have a significant impact on reducing greenhouse gas (GHG) emissions. For instance, individuals can reduce their carbon footprint by adopting energy-efficient practices, such as using LED light bulbs, improving home insulation, and utilizing energy-efficient appliances. Additionally, shifting to renewable energy sources, such as installing solar panels or choosing green energy plans, can further reduce emissions. Transportation is another area where individual actions can make a difference. Opting for public transport, carpooling, biking, or driving electric vehicles (EVs) instead of gasoline-powered cars can significantly reduce personal carbon emissions. Reducing, reusing, and recycling materials, as well as consuming less meat and processed foods, can also contribute to a lower carbon footprint. Moreover, advocating for stronger climate policies and supporting businesses that prioritize sustainability can amplify individual efforts. Collective action by individuals, when combined with governmental and industrial action, can help to reduce global emissions and accelerate the transition to a low-carbon future. By adopting more sustainable lifestyles, individuals can contribute to the global effort to combat climate change and create a more sustainable planet for future generations.

Carbon Pricing and Economic Solutions

Carbon pricing is a critical economic strategy for mitigating climate change by placing a price on carbon emissions, thereby incentivizing businesses and individuals to reduce their carbon footprint. The two most common approaches to carbon pricing are carbon taxes and cap-and-trade systems. A carbon tax directly imposes a fee on the carbon content of fossil fuels, encouraging companies to transition to cleaner energy sources by making carbon-intensive products more expensive. On the other hand, a cap-and-trade system sets a limit (cap) on the total amount of greenhouse gases (GHGs) that can be emitted and allows businesses to buy and sell emissions permits. This creates a financial incentive for companies to reduce their emissions to below the allocated cap, as they can sell their excess permits. Both systems aim to internalize the environmental costs of carbon emissions, ensuring that the price of fossil fuels reflects their true environmental impact. Carbon pricing not only helps reduce emissions but also generates revenue that can be reinvested in renewable energy projects, climate adaptation initiatives, and economic support for affected communities. Although carbon pricing is seen as a promising tool for climate mitigation, its effectiveness depends on proper implementation and political support. Challenges include setting the right price for carbon, ensuring equitable distribution of the costs and benefits, and addressing the potential for economic displacement in regions dependent on carbon-intensive industries. Despite these challenges, carbon pricing remains a key element in the transition to a low-carbon economy.

Climate Change Adaptation Strategies

Climate change adaptation refers to the process of making adjustments to social, economic, and environmental practices to minimize the negative impacts of climate change. While mitigation focuses on reducing greenhouse gas emissions to prevent further climate change, adaptation is concerned with responding to the changes that are already occurring. Effective adaptation strategies are crucial for reducing vulnerability and increasing resilience to climate change, particularly in regions that are already experiencing its effects. One key strategy is the development of resilient infrastructure that can withstand extreme weather events such as floods, storms, and heatwaves. For example, cities can invest in flood protection systems, resilient buildings, and cooling systems to combat rising temperatures. Agricultural adaptation strategies, such as the adoption of drought-resistant crops, improved irrigation techniques, and diversification of farming practices, can help ensure food security in areas experiencing changing rainfall patterns. Additionally, ecosystem-based adaptation, which involves protecting and restoring natural ecosystems such as mangroves, wetlands, and forests, can provide important services such as flood protection and carbon sequestration. Community-based adaptation strategies that empower local populations to engage in decision-making and prepare for climate risks are also crucial for building resilience. Adaptation must be inclusive and consider the needs of vulnerable populations, including low-income communities, indigenous peoples, and those in developing countries. Through coordinated efforts and investments in adaptation, societies can better cope with the unavoidable impacts of climate change.

The Role of Education in Climate Action

Education plays a pivotal role in combating climate change by raising awareness, building capacity, and fostering a sense of responsibility toward environmental sustainability. Climate education helps individuals understand the science of climate change, its causes and impacts, and the importance of taking action to reduce greenhouse gas emissions. Through education, people can be empowered to make informed decisions about their consumption patterns, lifestyle choices, and political engagement. Schools, universities, and community-based programs can play an essential role in fostering the next generation of climate leaders and activists. In addition to formal education, public awareness campaigns and media outreach can reach a broader audience and encourage collective action on climate issues. Climate change education can also support the development of skills and technologies that are critical for transitioning to a low-carbon economy. These include knowledge in renewable energy, sustainable agriculture, green building practices, and environmental policy. The United Nations Educational, Scientific and Cultural Organization (UNESCO) has emphasized the importance of integrating climate change education into national curricula to ensure that young people are equipped with the knowledge and skills necessary to face the challenges of climate change. By promoting climate literacy and empowering individuals, education contributes significantly to building a more sustainable and climate-resilient future.

Conclusion: A Path Forward

The challenge of climate change requires a multifaceted approach that involves the collective efforts of individuals, communities, governments, and industries worldwide. While the impacts of climate change are already being felt, there is still hope for mitigating its worst effects and adapting to the changes that are inevitable. Global cooperation is essential to implementing the necessary policy changes, from transitioning to renewable energy and adopting carbon pricing mechanisms to strengthening climate resilience and promoting sustainable development. Efforts to reduce greenhouse gas emissions must be intensified, and financial resources should be directed toward both mitigation and adaptation strategies, especially in vulnerable regions. The engagement of the private sector, civil society, and individuals is equally crucial in driving change at all levels. Climate change is not just an environmental issue but also a social, economic, and political one that affects every aspect of human life. By prioritizing climate action, we can create a sustainable future for generations to come. While the task is daunting, the solutions are within our reach, provided that we act decisively, boldly, and collectively. As we move forward, we must continue to work together, combining scientific innovation, policy solutions, and grassroots activism, to secure a world that is both resilient and sustainable for all.

Chapter 2: Pollution: Air, water, and land

Air Pollution: Causes and Consequences

Air pollution is one of the most significant environmental challenges facing the world today. It is caused by a variety of human activities, including industrial emissions, vehicle exhaust, agriculture, and the burning of fossil fuels. Air pollutants such as sulfur dioxide (SO2), nitrogen oxides (NOx), particulate matter (PM), and volatile organic compounds (VOCs) are released into the atmosphere, leading to harmful effects on both human health and the environment. Major sources of air pollution include power plants, factories, and transportation, especially in urban areas. These pollutants can lead to respiratory diseases, cardiovascular problems, and increased mortality rates. Children, the elderly, and individuals with pre-existing health conditions are particularly vulnerable to the effects of poor air quality. Moreover, air pollution contributes to environmental problems such as acid rain, which harms ecosystems and damages crops, as well as the formation of ground-level ozone, which leads to smog and reduced visibility. Long-term exposure to air pollution also contributes to climate change by increasing the concentration of greenhouse gases, particularly carbon dioxide (CO2) and methane (CH4). Addressing air pollution requires international cooperation, regulatory policies such as stricter emissions standards, and the promotion of cleaner technologies like electric vehicles and renewable energy. Additionally, raising public awareness and encouraging lifestyle changes, such as reducing car usage, can help mitigate air pollution and its harmful effects.

Water Pollution: Causes and Impact on Ecosystems

Water pollution is the contamination of water bodies, including rivers, lakes, and oceans, due to human activities and natural processes. The primary causes of water pollution include industrial waste, agricultural runoff, untreated sewage, and oil spills. These pollutants can lead to the degradation of water quality, making it unsafe for human consumption and aquatic life. Chemical pollutants, such as heavy metals, pesticides, and fertilizers, can enter water bodies through runoff from agricultural fields or improper waste disposal. As a result, aquatic ecosystems are disrupted, and species such as fish, aquatic plants, and microorganisms suffer. Additionally, waterborne diseases, caused by pathogens such as bacteria, viruses, and parasites, can spread through contaminated water, posing a serious public health risk. In many developing countries, inadequate access to clean water and sanitation infrastructure exacerbates the problem, leading to widespread waterborne diseases. The accumulation of plastic waste in oceans has also become a major concern, as marine animals ingest or become entangled in plastic debris, resulting in injury or death. The depletion of freshwater resources due to over-extraction further exacerbates the issue, particularly in arid regions. To combat water pollution, governments must implement stringent water quality standards, promote sustainable agricultural practices, and invest in wastewater treatment facilities. Public awareness campaigns, alongside technological innovations in water purification and recycling, can also contribute to preserving water resources for future generations.

Land Pollution: Causes, Effects, and Solutions

Land pollution refers to the degradation of the Earth's surface caused by the accumulation of waste materials and harmful chemicals. It is primarily the result of human activities such as improper waste disposal, deforestation, mining, and industrial activities. Solid waste, including plastics, metals, and electronic waste, constitutes a major component of land pollution. The improper disposal of these materials in landfills, or their dumping in open areas, leads to contamination of soil and groundwater

(Environmental Protection Agency [EPA], 2020). The accumulation of non-biodegradable waste, especially plastics, poses a long-term environmental threat, as these materials can remain in the soil for hundreds of years. In addition, the use of toxic chemicals in farming, such as pesticides and herbicides, can lead to soil degradation and contamination, making the land unsuitable for agriculture. Land pollution also results from urban sprawl, where natural habitats are replaced by concrete structures, leading to the loss of biodiversity and the disruption of ecosystems. The impact of land pollution on human health is significant, as contaminated soil can lead to the spread of diseases and pose a direct threat to food safety. To address land pollution, governments and local authorities must implement waste management systems that encourage recycling, reduce waste generation, and promote composting. Educating the public about responsible waste disposal and encouraging the reduction of single-use plastics are also critical in minimizing land pollution. Furthermore, reforestation and soil conservation efforts can help restore degraded land and prevent further damage.

Integrated Solutions to Pollution

Addressing pollution in its various forms requires a multi-faceted approach that combines regulation, innovation, and community involvement. Governments must establish and enforce policies that set standards for air, water, and land quality, while industries must adopt cleaner technologies and practices to reduce emissions and waste. Environmental education plays a key role in raising public awareness and encouraging sustainable behavior. For instance, individuals can contribute to pollution reduction by reducing their carbon footprint, conserving water, and recycling waste. On a larger scale, cities can invest in green infrastructure, such as renewable energy systems, waste-to-energy plants, and public transportation, to reduce the impact of urbanization. Technology, such as advanced water treatment plants, air filtration systems, and waste management technologies, can significantly mitigate pollution's adverse effects. Moreover, international cooperation is vital, as pollution does not respect borders. Global agreements like the Paris Agreement on climate change and the UN's Sustainable Development Goals (SDGs) provide frameworks for collective action on environmental issues. To achieve meaningful progress in combating pollution, there must be a commitment to sustainability at all levels of society—from governments and businesses to individuals.

Air Pollution and Its Impact on Public Health

Air pollution is a major public health issue that affects millions of people worldwide. Exposure to polluted air can cause a wide range of health problems, from respiratory diseases like asthma to cardiovascular conditions and even premature death. Fine particulate matter (PM2.5), which is small enough to penetrate deep into the lungs and bloodstream, is particularly dangerous. Studies have shown that long-term exposure to high levels of air pollution increases the risk of chronic respiratory diseases, heart attacks, strokes, and lung cancer. In addition to physical health impacts, air pollution can also contribute to mental health problems, such as depression and anxiety, especially in urban areas with high pollution levels. Vulnerable populations, including children, the elderly, and individuals with pre-existing health conditions, are at greater risk. For example, children exposed to polluted air are more likely to experience developmental issues and reduced lung function. Furthermore, air pollution disproportionately affects low-income communities, where access to healthcare and clean air is often limited. Addressing the health impacts of air pollution requires a comprehensive approach that includes stricter emissions standards, the promotion of clean energy technologies, and public health campaigns to raise awareness. Reducing air pollution not only improves the quality of life but also decreases the burden on healthcare systems, ultimately saving lives and reducing economic costs.

Water Pollution and Its Threat to Biodiversity

Water pollution is a critical threat to biodiversity, particularly in aquatic ecosystems. Contaminants such as industrial chemicals, agricultural runoff, and untreated sewage can harm aquatic life by disrupting ecosystems and food chains. For example, nitrogen and phosphorus from fertilizers cause nutrient pollution, leading to algal blooms that deplete oxygen levels in water, resulting in "dead zones" where most marine life cannot survive. Heavy metals like mercury and lead are toxic to fish and other aquatic organisms, causing deformities, reproductive failure, and even death. Furthermore, plastics and microplastics have become pervasive pollutants in oceans, rivers, and lakes, harming wildlife through ingestion or entanglement. Marine species, including sea turtles, whales, and seabirds, often mistake plastic debris for food, leading to injuries or death. Biodiversity loss in aquatic ecosystems has far-reaching consequences for human populations that rely on these ecosystems for food, livelihoods, and water resources. The depletion of fish stocks and the degradation of water quality can have economic and social implications for communities dependent on fisheries and tourism. To protect biodiversity and preserve water resources, it is essential to implement stronger regulations on industrial discharges, improve wastewater treatment systems, and reduce plastic use. Conservation efforts that protect aquatic habitats, such as wetlands and coral reefs, are also vital in maintaining biodiversity.

Land Pollution and Its Economic Costs

Land pollution is not only an environmental issue but also a significant economic problem. The improper disposal of waste, including plastics, e-waste, and industrial by-products, results in land degradation that incurs high costs for society. Contaminated land requires extensive cleanup efforts, which often involve expensive remediation techniques, such as soil washing, chemical treatments, and landfill containment. In urban areas, land pollution reduces property values and hinders economic development, as polluted land becomes less attractive for investment or residential use. Additionally, landfills, which are often the final destination for much of the world's waste, take up large areas of land, creating a strain on available real estate. Land pollution can also reduce agricultural productivity by contaminating soil and water sources, which can result in reduced crop yields and higher food prices. The economic costs of land pollution are often passed on to communities through increased healthcare expenses due to exposure to hazardous chemicals and diseases caused by contaminated environments. To address the economic impacts of land pollution, governments must implement effective waste management systems, promote recycling, and encourage businesses to adopt sustainable production practices. Investing in land restoration projects and transitioning to circular economies can help reduce the economic burden of land pollution and promote long-term environmental sustainability.

The Role of Policy in Reducing Pollution

Effective policy is essential in addressing the global pollution crisis. Governments play a key role in setting regulations and enforcing standards that limit the release of pollutants into the air, water, and land. One of the most important policy tools for reducing pollution is the establishment of environmental regulations, such as the Clean Air Act and Clean Water Act in the United States, which set limits on emissions and discharges from industries. International agreements, such as the Paris Agreement on climate change, provide frameworks for reducing pollution on a global scale by setting emission reduction targets for countries. In addition to regulatory measures, governments can incentivize the adoption of green technologies and renewable energy sources by offering subsidies, tax credits, and grants. The promotion of sustainable agriculture, waste management, and transportation systems can further reduce the environmental impact of human activities. Furthermore, strong public-private partnerships are crucial for driving innovation in pollution control technologies, such as advanced air filtration systems, wastewater treatment, and waste-to-energy technologies. Public policy can also encourage individual behavioral changes, such as promoting the use of public transport, reducing plastic consumption, and encouraging recycling. Ultimately, political will, effective governance, and strong enforcement of environmental laws are necessary to combat pollution and protect human and environmental health.

The Link Between Pollution and Climate Change

Pollution is intrinsically linked to climate change, as many pollutants contribute to the warming of the planet. Air pollution, in the form of greenhouse gases such as carbon dioxide (CO2), methane (CH4), and nitrous oxide (N2O), traps heat in the Earth's atmosphere, leading to global warming and climate disruptions. The burning of fossil fuels for energy production, transportation, and industrial processes is a major source of these pollutants. In addition to greenhouse gases, black carbon (a component of particulate matter) also contributes to warming by absorbing sunlight and accelerating ice melt in polar regions. Water pollution exacerbates the effects of climate change, as polluted water bodies are less able to absorb carbon dioxide, leading to further atmospheric buildup of greenhouse gases. In some cases, pollution-induced environmental damage, such as deforestation and the destruction of wetlands, can reduce the planet's natural ability to regulate the climate. For instance, the clearing of forests for agriculture releases stored carbon into the atmosphere, contributing to both pollution and climate change. Addressing the pollution-climate change nexus requires integrated approaches that target both the reduction of pollutants and the mitigation of climate change. This includes transitioning to renewable energy sources, improving waste management, and adopting sustainable land use practices that reduce emissions and protect ecosystems.

Industrial Pollution: Sources and Solutions

Industrial pollution is a significant contributor to air, water, and land pollution, as manufacturing processes release various harmful substances into the environment. Factories, power plants, and other industrial facilities emit pollutants such as sulfur dioxide (SO2), nitrogen oxides (NOx), carbon monoxide (CO), heavy metals, and hazardous chemicals. Air pollution from industries can lead to smog formation and acid rain, which damages ecosystems, crops, and buildings. Water pollution arises when industries discharge untreated waste, including toxic chemicals and heavy metals, into rivers and lakes, causing harm to aquatic life and making water unsafe for human consumption. Land pollution from industrial waste includes the accumulation of hazardous materials, which can contaminate the soil and make land unfit for agriculture. Industrial pollution can also contribute to global climate change through the emission of greenhouse gases, particularly CO2 from energy-intensive processes. Addressing industrial pollution requires stricter regulations on emissions, mandatory waste treatment technologies, and the promotion of cleaner production methods. For example, industries can adopt eco-friendly technologies, such as green chemistry, renewable energy sources, and closed-loop recycling, to reduce their environmental footprint. Additionally, governments must enforce laws that penalize non-compliance with environmental standards and incentivize the adoption of sustainable practices through tax breaks and subsidies for green innovations.

Agricultural Pollution: Impact on Water and Soil Quality

Agricultural practices are a leading source of water and land pollution, as the use of chemical fertilizers, pesticides, and herbicides can degrade water quality and harm soil health. Fertilizers containing nitrogen and phosphorus are often applied to boost crop yields, but excessive use leads to nutrient pollution in nearby water bodies. Runoff from agricultural fields carries these nutrients into rivers and lakes, where they cause eutrophication, promoting harmful algal blooms and creating "dead zones" where oxygen levels are too low to support most aquatic life. Additionally, pesticides and herbicides can contaminate both water and soil, posing risks to biodiversity and human health. Over time, these chemicals can accumulate in the environment, leading to long-term environmental damage and resistance in pests. Agricultural runoff also carries pathogens and pathogens from animal waste, contributing to the spread of waterborne diseases. To mitigate agricultural pollution, sustainable farming practices must be adopted, such as integrated pest management, crop rotation, and organic farming. Farmers can also implement best management practices, such as buffer zones around water bodies, precision farming techniques, and the use of organic fertilizers to reduce chemical inputs. These methods help protect soil and water quality while maintaining agricultural productivity and reducing the environmental footprint.

Plastic Pollution: A Global Crisis

Plastic pollution is one of the most pervasive environmental problems of the modern era, affecting land, water, and air. Plastics are non-biodegradable materials that persist in the environment for hundreds of years, accumulating in landfills, oceans, rivers, and other ecosystems. The excessive use of plastic products, particularly single-use plastics like bottles, bags, and packaging, has led to a global crisis of plastic waste (National Geographic, 2020). In oceans, plastics pose a direct threat to marine life, as animals mistake plastic debris for food or become entangled in it, leading to injury or death. Plastics also break down into smaller particles known as microplastics, which can be ingested by marine organisms, entering the food chain and potentially affecting human health. On land, plastic waste clogs drainage systems, disrupts ecosystems, and contributes to land pollution. The widespread use of plastics is driven by their low cost, convenience, and versatility, but these advantages come at a steep environmental cost. Solutions to plastic pollution include reducing plastic consumption through reusable products, improving waste management systems, and promoting recycling and upcycling. Bans on single-use plastics, such as plastic bags and straws, have been adopted in some countries, while others have implemented extended producer responsibility (EPR) programs that hold manufacturers accountable for the lifecycle of their products. Additionally, innovations in biodegradable plastics and waste-to-energy technologies offer promising alternatives.

The Role of Technology in Combatting Pollution

Technology plays a vital role in combating pollution by providing innovative solutions to reduce emissions, treat waste, and monitor environmental quality. Advances in renewable energy technologies, such as solar, wind, and hydropower, have enabled a shift away from fossil fuels, significantly reducing air pollution and greenhouse gas emissions. Similarly, the development of electric vehicles (EVs) is helping to reduce air pollution in urban areas, as EVs produce zero tailpipe emissions. In water treatment, technologies like reverse osmosis, filtration systems, and bio-remediation techniques are being used to remove pollutants from wastewater and restore clean water supplies. In terms of land pollution, waste-to-energy technologies allow for the conversion of waste materials into usable energy, reducing the burden on landfills and generating clean power. Moreover, environmental monitoring technologies, such as air quality sensors and satellite imaging, enable real-time tracking of pollution levels and provide valuable data for policy makers to take proactive measures. While technological solutions are crucial in addressing pollution, they must be accompanied by strong regulatory frameworks, public awareness, and investment in sustainable practices. Governments and industries must collaborate to ensure that these technologies are accessible and effective in tackling pollution on a global scale.

Waste Management: Key to Reducing Land Pollution

Waste management is an essential component in reducing land pollution. The improper disposal of solid waste, including household, industrial, and agricultural waste, leads to the contamination of land, water, and air. Landfills, which are the most common disposal method, often release harmful gases like methane and leachate into the soil and groundwater, posing environmental and health risks. To address these challenges, effective waste management strategies must be implemented. This includes reducing waste generation through the promotion of reduce, reuse, and recycle practices, as well as improving recycling programs to recover valuable materials and reduce landfill usage. Waste segregation at the source allows for better processing and recycling, ensuring that recyclable materials like paper, plastics, and metals are diverted from landfills. In addition, composting organic waste can help reduce methane emissions and provide valuable fertilizers for agriculture. Another important aspect of waste management is waste-to-energy technologies, which convert waste materials into usable energy, reducing the reliance on landfills and incinerators. Proper disposal of hazardous waste is also crucial to prevent soil contamination and the spread of toxins. Governments must implement regulations and incentives that encourage responsible waste disposal, while individuals and businesses must play an active role in reducing waste and adopting sustainable practices.

The Impact of Deforestation on Pollution

Deforestation is a major driver of pollution, particularly in terms of land degradation and climate change. Trees play a critical role in maintaining soil quality, filtering air pollutants, and sequestering carbon dioxide (CO2). However, large-scale deforestation, often for agricultural expansion, logging, and urbanization, disrupts these functions and leads to a variety of environmental issues. When forests are cleared, the carbon stored in trees is released into the atmosphere, contributing to global warming. Deforestation also leads to soil erosion, as trees and vegetation are vital in stabilizing soil and preventing runoff. Without tree cover, the soil becomes more prone to erosion, which can lead to the loss of fertile land, increased sedimentation in rivers, and poor water quality. Furthermore, deforestation reduces biodiversity, as it destroys habitats for countless species. The loss of forests not only harms the environment but also affects human populations, particularly those who depend on forests for livelihoods and resources. To combat the impact of deforestation, countries must implement sustainable land-use practices, including reforestation and afforestation efforts, and promote forest conservation. Governments, NGOs, and local communities must collaborate to protect existing forests and ensure that new development is done in an environmentally responsible manner.

Ocean Pollution: Effects on Marine Life and Human Health

Ocean pollution is a growing global concern, as it impacts marine ecosystems and the species that rely on them, including humans. Major pollutants in the ocean include plastics, oil, chemicals, and agricultural runoff. Plastics are particularly problematic, as they are non-biodegradable and accumulate in the ocean, creating large debris fields known as "garbage patches." Marine animals often ingest plastic or become entangled in it, leading to injury or death. The ingestion of plastics by marine species also poses risks to human health, as seafood can carry microplastics, which have been linked to various health issues, including gastrointestinal problems and hormonal disruption (National Oceanic and Atmospheric Administration [NOAA], 2021). Oil spills, often caused by shipping accidents or offshore drilling, can devastate marine ecosystems by coating fish, birds, and other wildlife in toxic substances, reducing their ability to survive and reproduce. Agricultural runoff, which carries pesticides, fertilizers, and sediments into the ocean, can lead to nutrient pollution, causing algal blooms that deplete oxygen and create "dead zones." To address ocean pollution, global efforts must focus on reducing plastic waste, improving waste management practices, and enforcing stricter regulations on pollution from shipping and industrial activities. Public education campaigns, along with innovations in waste disposal and recycling technologies, are also essential for reducing ocean pollution.

The Role of Urbanization in Pollution

Urbanization plays a significant role in increasing pollution levels, particularly air, water, and land pollution. As cities expand and populations grow, the demand for energy, transportation, and goods increases, leading to higher emissions from vehicles, industrial activities, and energy production. Urban areas often experience higher concentrations of air pollutants, such as carbon monoxide (CO), nitrogen oxides (NOx), and particulate matter, due to heavy traffic and industrial emissions. Water pollution in cities

The Role of Urbanization in Pollution

Water pollution in cities occurs due to the runoff of pollutants from roads, construction sites, and untreated sewage, which contaminates rivers, lakes, and groundwater. The high density of buildings and infrastructure reduces natural areas that can absorb rainwater, leading to increased stormwater runoff. This results in the contamination of water sources with harmful chemicals and waste. Furthermore, land pollution is common in urban environments due to improper waste disposal, with landfills and illegal dumping sites contributing to soil contamination and degradation. Urbanization also places a significant strain on public health, as air and water pollution in cities are linked to various diseases, including respiratory problems, heart disease, and waterborne diseases. The development of green spaces, improved public transportation, and sustainable waste management practices can help mitigate the negative environmental impacts of urbanization. Cities can reduce their pollution footprint by adopting eco-friendly building materials, investing in renewable energy sources, and promoting urban agriculture. Additionally, urban planning policies that integrate green spaces, reduce traffic congestion, and improve waste infrastructure can help create healthier urban environments. By focusing on sustainable development, urban areas can strike a balance between growth and environmental protection.

Noise Pollution: An Overlooked Environmental Issue

Noise pollution, although often overlooked, is an important environmental issue with significant effects on both human health and wildlife. Urbanization, industrialization, and transportation systems contribute to high levels of noise in many areas, affecting communities and ecosystems. Long-term exposure to elevated noise levels is associated with various health problems, such as hearing impairment, sleep disturbances, increased stress levels, and cardiovascular diseases. In wildlife, noise pollution disrupts communication, breeding patterns, and migration, which can lead to long-term population declines. For example, animals that rely on echolocation, such as bats and dolphins, are particularly vulnerable to the interference of loud sounds from vehicles, ships, or construction sites. Additionally, excessive noise can disturb the natural balance of ecosystems by altering predator-prey relationships and disrupting feeding or mating behaviors. To address noise pollution, urban planners must incorporate noise-reducing measures in infrastructure projects, such as using noise barriers, reducing traffic congestion, and adopting quieter transportation technologies. Regulations on industrial noise, soundproofing in residential areas, and public awareness campaigns can also help minimize the effects of noise pollution. As communities continue to grow, it is essential to prioritize the health and well-being of individuals and wildlife by minimizing exposure to harmful noise levels.

The Economic Costs of Pollution

Pollution has significant economic implications, as it contributes to higher healthcare costs, reduced agricultural productivity, and lower quality of life. The healthcare costs associated with pollution-related diseases, such as respiratory illnesses, cardiovascular diseases, and cancer, place a heavy burden on public health systems. For example, the World Health Organization (WHO) estimates that outdoor air pollution costs the global economy approximately $5 trillion annually in lost productivity and healthcare expense. In addition, pollution negatively affects agricultural productivity by contaminating soil, water, and air, which can lead to reduced crop yields, loss of arable land, and increased food prices. The degradation of ecosystems and the loss of biodiversity also have long-term economic consequences, as healthy ecosystems provide essential services such as clean water, soil fertility, and pollination. Furthermore, pollution can reduce tourism revenue, as polluted environments are less attractive to visitors. Governments can mitigate the economic costs of pollution by implementing environmental policies that promote cleaner industries, renewable energy, and sustainable agriculture. By investing in pollution control technologies, promoting environmental conservation, and encouraging green jobs, governments can reduce the economic burden of pollution while creating a more sustainable economy.

Global Cooperation in Addressing Pollution

Addressing the global pollution crisis requires coordinated action at the international level. Pollution knows no borders, and many environmental problems, such as air pollution and climate change, have transboundary effects. Therefore, countries must work together to develop and implement global strategies to reduce pollution and protect the environment. International agreements, such as the Paris Agreement on climate change, play a crucial role in uniting countries in the fight against pollution and climate-related environmental challenges. These agreements establish common goals and commitments for reducing greenhouse gas emissions, transitioning to renewable energy, and implementing climate adaptation strategies. Furthermore, global cooperation is essential in addressing the issue of plastic pollution, with initiatives like the United Nations' Clean Seas Campaign aiming to reduce plastic waste and promote sustainable consumption and production practices. Collaborative efforts also involve sharing knowledge, technology, and best practices in pollution control, as well as providing financial support to developing countries that may lack the resources to combat pollution effectively. Multilateral organizations, non-governmental organizations, and the private sector also play an important role in addressing pollution by raising awareness, funding research, and developing innovative solutions. A global approach to pollution is necessary to ensure a cleaner, healthier, and more sustainable planet for future generations.

Chapter 3: Deforestation: Impact on ecosystems

Introduction to Deforestation and Its Environmental Impacts

Deforestation is one of the leading environmental issues facing the world today. It involves the large-scale clearing of forests, primarily for agricultural, industrial, or urban development purposes. Forests are crucial ecosystems that support a wide variety of species and regulate critical natural processes, such as water cycles, carbon storage, and soil fertility. When forests are removed, the consequences for biodiversity and climate regulation are profound. In addition to the loss of biodiversity, deforestation disrupts vital ecological functions that support human life. According to the United Nations (2019), over 10 million hectares of forests are lost every year, contributing to the degradation of ecosystems worldwide. Forests act as carbon sinks, absorbing carbon dioxide (CO2) from the atmosphere, and their destruction releases carbon back into the atmosphere, exacerbating climate change. The impacts of deforestation are not limited to environmental degradation, as it also affects local economies and the livelihoods of millions of people who depend on forests for sustenance, medicine, and shelter. This paper will explore the effects of deforestation on ecosystems, including biodiversity loss, soil degradation, and climate change, while highlighting potential solutions for mitigating these impacts.

Causes of Deforestation

The primary causes of deforestation can be categorized into both direct and indirect drivers. Direct causes include logging for timber, agriculture (especially commercial agriculture for crops like palm oil, soy, and cocoa), and urbanization. Logging involves the removal of trees for timber and paper products, while agricultural expansion leads to the clearing of forests to create farmland or pastures for livestock. According to the World Resources Institute (2020), agricultural expansion is responsible for nearly 80% of deforestation globally. The demand for palm oil, for instance, has led to widespread deforestation in Southeast Asia, particularly in Indonesia and Malaysia. Indirect causes, on the other hand, stem from economic policies, weak governance, and unsustainable land-use practices. In many developing countries, poor land tenure systems and inadequate enforcement of environmental laws contribute to illegal deforestation. Moreover, economic incentives that prioritize short-term profit over environmental protection, such as subsidies for unsustainable farming, further exacerbate deforestation. These causes of deforestation are interlinked, creating a complex web of factors that drive forest loss on a global scale. To effectively combat deforestation, it is crucial to address both direct and indirect causes through comprehensive policy and regulatory frameworks.

Deforestation and Its Impact on Biodiversity

Forests are home to more than half of the world's terrestrial species, making deforestation a significant threat to biodiversity. The loss of forests results in habitat destruction, forcing many species to either relocate or face extinction. Deforestation is particularly detrimental to endemic species, which are only found in specific areas and cannot survive outside their natural habitats. For example, the deforestation of the Amazon rainforest has contributed to the endangerment of species such as the jaguar and the Amazon River dolphin. Additionally, the fragmentation of forests, which occurs when large forests are broken into smaller patches, isolates wildlife populations, limiting their ability to find food, mates, and migrate. This isolation can lead to genetic bottlenecks, reducing genetic diversity and making species more vulnerable to disease and environmental changes. Deforestation also exposes species to increased predation and competition from invasive species, further threatening native populations. To preserve biodiversity, it is essential to protect large, contiguous forest areas, restore fragmented ecosystems, and reduce the rates of deforestation through sustainable land-use practices and conservation efforts.

Soil Degradation and Erosion Due to Deforestation

Deforestation contributes significantly to soil degradation and erosion. Trees and vegetation play a vital role in preventing soil erosion by stabilizing the soil with their roots, which bind the soil particles together. When forests are cleared, the soil becomes vulnerable to erosion by wind and water. The removal of trees also reduces the amount of organic matter in the soil, which is essential for maintaining soil fertility. In tropical regions, where soils are often nutrient-poor, deforestation leads to a rapid decline in soil quality, making it difficult for crops to thrive. The loss of soil fertility also increases the need for chemical fertilizers, which can further degrade the soil and pollute nearby water bodies. Moreover, erosion leads to sedimentation in rivers and lakes, reducing water quality and affecting aquatic life. In areas prone to deforestation, such as the Amazon rainforest, soil erosion and degradation pose significant challenges to sustainable agriculture and food security. To prevent these outcomes, it is crucial to adopt sustainable land-management practices that protect the soil, such as agroforestry, contour farming, and reforestation.

Deforestation and the Water Cycle

Forests play a crucial role in regulating the water cycle, and their destruction disrupts this balance. Trees absorb water from the soil and release it into the atmosphere through transpiration, which contributes to cloud formation and precipitation (Ellis et al., 2010). When forests are cleared, the moisture released into the atmosphere decreases, which can lead to reduced rainfall and droughts. In tropical regions, such as the Amazon, deforestation has been linked to a reduction in regional rainfall, affecting local agriculture and water availability (Zhao et al., 2019). Furthermore, the removal of forests increases surface runoff, as there are fewer tree roots to absorb water. This leads to soil erosion, sedimentation in rivers, and the degradation of water quality. The disruption of the water cycle also affects the global climate, as changes in rainfall patterns can lead to altered weather patterns, including more frequent storms and droughts. To mitigate these impacts, it is essential to protect forests, promote reforestation, and implement water-conservation practices.

Deforestation and Climate Change

Deforestation is a major driver of climate change due to the significant amounts of carbon dioxide (CO2) released into the atmosphere when forests are destroyed. Trees act as carbon sinks, absorbing CO2 during photosynthesis and storing carbon in their biomass. When forests are cleared, the carbon stored in the trees is released back into the atmosphere, contributing to global warming. Deforestation accounts for approximately 10% of global greenhouse gas emissions, making it one of the largest anthropogenic sources of CO2 emissions. The effects of deforestation on climate change are particularly severe in tropical regions, where the rapid growth of forests has a significant role in sequestering carbon. As forests are replaced by agricultural land or urban areas, not only is the carbon sequestration capacity reduced, but the release of CO2 from the decaying vegetation and soil further exacerbates climate change. Additionally, deforestation impacts the Earth's albedo, the reflectivity of the surface, which can alter weather patterns and increase global temperatures. To mitigate these effects, it is crucial to reduce deforestation and promote reforestation and afforestation efforts, which can enhance carbon storage and help regulate global climate patterns.

Deforestation and Its Effect on Indigenous Communities

Indigenous communities have long depended on forests for their livelihoods, culture, and survival. Deforestation often disproportionately impacts these communities, who rely on the forest for food, medicine, shelter, and spiritual practices. When forests are destroyed, indigenous people lose not only their homes and resources but also their connection to the land and cultural identity. In many parts of the world, indigenous communities are the first to bear the brunt of deforestation, especially when large corporations or governments clear forests for agriculture, mining, or infrastructure development. For instance, in the Amazon, the expansion of cattle ranching and soy plantations has led to the displacement of numerous indigenous groups. These communities often face violence, land rights violations, and social conflicts as a result of deforestation. Furthermore, deforestation leads to the degradation of traditional farming systems, which depend on healthy forests for the regulation of water, soil fertility, and biodiversity. To address these issues, it is essential to respect the land rights of indigenous peoples and involve them in decision-making processes related to forest management and conservation. Protecting indigenous communities' rights to land and resources can be a powerful tool in halting deforestation and promoting sustainable forest management.

The Economic Impacts of Deforestation

Deforestation has a significant economic impact, not only on the environment but also on the economy. While the immediate benefits of deforestation, such as timber extraction, agricultural expansion, and urbanization, can seem lucrative, the long-term costs are far more severe. Deforestation disrupts ecosystem services, such as water regulation, soil fertility, and climate regulation, all of which are crucial for sustainable economic development. For instance, the degradation of soil caused by deforestation reduces agricultural productivity, requiring farmers to rely more on expensive fertilizers and irrigation. Additionally, the loss of forests reduces biodiversity, which has direct economic consequences for industries such as pharmaceuticals, agriculture, and tourism. The Amazon rainforest, for example, is a critical source of medicinal plants, many of which have not yet been studied for their potential benefits. Furthermore, deforestation accelerates climate change, which results in more frequent and severe natural disasters, such as floods, droughts, and wildfires. These disasters can cause significant damage to infrastructure, disrupt local economies, and result in high recovery costs. To ensure long-term economic stability, it is essential to recognize the value of forest ecosystems and integrate sustainable practices into development strategies.

Deforestation and Soil Carbon Release

The clearing of forests releases vast amounts of carbon stored in both vegetation and soils into the atmosphere, contributing to global warming. Forests store significant amounts of carbon in their biomass and soil, and when these forests are cleared, the carbon is released through decomposition, burning, and soil disturbance. Deforestation contributes nearly 10% of global greenhouse gas emissions. This process is particularly severe in tropical forests, where the carbon content in the soil is higher due to the rapid decay of organic material. In addition to releasing carbon, deforestation can cause a reduction in the soil's ability to store carbon in the future. Soil degradation and erosion that often accompany deforestation further diminish the soil's capacity to act as a carbon sink (Lal, 2004). The loss of forest ecosystems and their carbon sequestration potential represents a significant threat to efforts aimed at mitigating climate change. To address these challenges, it is crucial to focus on forest conservation and reforestation efforts, as well as the sustainable management of soil and land resources.

Deforestation and the Risk of Wildfires

Deforestation increases the frequency and intensity of wildfires, particularly in regions that are prone to dry conditions. Forests play an important role in regulating fire behavior by maintaining moisture levels in the environment and providing a natural buffer against fire spread (Pechony & Shindell, 2010). When forests are cleared, the vegetation that would typically act as a firebreak is removed, and dry conditions are exacerbated by the loss of tree cover. This creates an environment more conducive to wildfires. For instance, in areas like the Amazon, deforestation is often followed by intentional fires set to clear land for agriculture (Aragão et al., 2018). These fires can spread uncontrollably, resulting in catastrophic destruction of both forests and nearby communities. Additionally, wildfires release large amounts of CO2 into the atmosphere, further exacerbating climate change. The increase in wildfires due to deforestation underscores the need for forest management strategies that incorporate fire prevention and mitigation measures. Promoting sustainable land use, protecting forests, and implementing firebreaks are essential steps in reducing the risk of wildfires.

Deforestation and Its Effects on Global Water Systems

Deforestation significantly disrupts global water systems, which are essential for maintaining the health of ecosystems and human populations. Forests play a critical role in the water cycle by regulating the flow of water into rivers and lakes. They act as natural filters, absorbing rainfall and allowing it to gradually infiltrate the ground. In turn, this helps to maintain groundwater levels and ensures a steady flow of freshwater in rivers. When forests are cleared, however, this natural filtration system is lost. This leads to increased surface runoff, which not only causes soil erosion but also results in the contamination of water bodies with sediments, pollutants, and nutrients. In tropical regions, deforestation has been linked to reduced rainfall, droughts, and disruptions in seasonal water availability. For example, the loss of forests in the Amazon basin is associated with significant changes in regional hydrological cycles, affecting water resources for millions of people. To prevent further disruption of global water systems, it is essential to protect forest ecosystems, implement sustainable land use practices, and restore damaged areas.

Deforestation and Its Impact on Carbon Sequestration

Carbon sequestration is one of the critical ecosystem services provided by forests, where trees absorb carbon dioxide (CO2) and store it in their biomass and soil. Deforestation disrupts this natural carbon storage process, leading to a significant release of carbon into the atmosphere, which contributes to global warming. Tropical forests, such as the Amazon, are particularly important carbon sinks, storing approximately one-quarter of the Earth's terrestrial carbon. When forests are cleared, the carbon stored in both the trees and soil is released into the atmosphere, increasing greenhouse gas concentrations and accelerating climate change. Furthermore, deforestation reduces the overall capacity of the Earth to absorb CO2, which exacerbates the global carbon cycle imbalance. The loss of forests also leads to a reduction in the carbon sequestration potential of the land, as the soil's ability to store carbon diminishes when forest cover is removed. To mitigate climate change, protecting existing forests and reforesting degraded areas is crucial in enhancing carbon sequestration capacity.

Deforestation and Human Health

The impacts of deforestation extend beyond environmental consequences to human health. Forests provide critical ecosystem services, including air and water purification, temperature regulation, and the prevention of vector-borne diseases. Deforestation disrupts these services, often resulting in negative health outcomes. For example, forests help regulate the flow of water, reducing the occurrence of floods and maintaining clean water sources for communities. Without forests, increased runoff leads to water contamination, posing risks to public health. Additionally, the destruction of forests can contribute to the spread of infectious diseases. Research has shown that deforestation and habitat fragmentation increase the likelihood of human-wildlife interactions, which can lead to the transmission of diseases such as malaria and dengue fever (Patz et al., 2004). The loss of forest biodiversity also means the loss of potential medicinal resources, as forests are a rich source of plants used in traditional medicine. In tropical regions, deforestation is associated with rising health risks due to increased exposure to pathogens and pollutants. To protect human health, it is essential to safeguard forests and promote sustainable land use that prioritizes both environmental and public health.

Deforestation and the Loss of Ecosystem Services

Ecosystem services are the benefits that humans derive from ecosystems, including forests. These services include food, water, air purification, climate regulation, and soil fertility. Deforestation has significant consequences for these services, which are essential for human survival. Forests provide a wide range of ecosystem services, such as pollination, carbon sequestration, and water regulation. The loss of forests leads to the depletion of these services, resulting in negative impacts on agriculture, water supply, and climate stability. For example, forests regulate the water cycle by absorbing rainwater and releasing it into the atmosphere, which helps maintain local and regional rainfall patterns. Without forests, the disruption of this cycle can lead to droughts and water shortages, affecting millions of people. Additionally, the loss of forests reduces biodiversity, which has direct economic and cultural implications for societies that rely on forest resources for food, medicine, and income. The depletion of ecosystem services further exacerbates poverty and inequality, particularly for indigenous communities that depend on forests for their livelihoods. To address these issues, it is essential to integrate the value of ecosystem services into policy-making and land-use planning.

Deforestation and Its Effects on Global Food Security

Forests play an important role in supporting global food security by maintaining soil fertility, regulating the water cycle, and providing direct food resources such as fruits, nuts, and animals. The loss of forests, however, can negatively impact food security by reducing agricultural productivity and increasing the risks of crop failure. Deforestation leads to soil degradation and erosion, which reduces soil fertility and makes it more difficult to grow crops. In many tropical regions, such as Southeast Asia and the Amazon, deforestation has resulted in the loss of fertile soils that are essential for agriculture. Furthermore, the loss of forests disrupts the water cycle, leading to droughts, floods, and irregular rainfall patterns, which further threaten food production. The destruction of forests also impacts smallholder farmers and indigenous communities who rely on forests for food, medicine, and other resources. By protecting forests and promoting sustainable land-use practices, such as agroforestry and organic farming, food security can be enhanced, and the negative impacts of deforestation on agriculture can be mitigated.

Deforestation and the Alteration of Local Climates

Deforestation leads to significant alterations in local and regional climates. Trees and forests are key components of the Earth's climate system, providing critical services such as shade, moisture retention, and carbon storage. When forests are removed, the environment undergoes rapid changes, including increases in temperature and changes in precipitation patterns. Forests regulate temperature by releasing moisture into the air through transpiration and by providing shade, both of which help moderate local climates. Without these natural buffers, regions can experience more extreme temperature fluctuations, leading to hotter, drier conditions. For instance, in areas where tropical rainforests are cleared, the absence of canopy cover leads to increased ground temperatures and reduced humidity, which exacerbates drought conditions. Moreover, deforestation contributes to altered precipitation patterns, often resulting in reduced rainfall and more frequent or severe droughts. This disruption in the local water cycle can impact agriculture, water supply, and overall ecosystem health. To mitigate these effects, forest conservation and reforestation are essential in maintaining local climate stability, enhancing resilience to climate change, and protecting the livelihoods of affected communities.

Deforestation and Its Role in Biodiversity Loss

Deforestation is one of the primary drivers of biodiversity loss around the world. Forests are home to a wide range of species, many of which are endemic and cannot survive outside their natural habitat. When forests are destroyed or fragmented, these species are often left without the resources they need to thrive. For example, the Amazon rainforest, home to more than 10% of the world's biodiversity, is being rapidly deforested for agriculture, logging, and urban expansion, putting countless species at risk of extinction. Deforestation also leads to habitat fragmentation, which isolates animal populations and makes it difficult for them to find food, mates, and shelter. In many cases, this fragmentation results in the loss of genetic diversity, which weakens the resilience of species to disease and environmental stress. Furthermore, the loss of biodiversity impacts ecosystem functioning, as many species play critical roles in pollination, seed dispersal, and nutrient cycling (Cardinale et al., 2012). To prevent further biodiversity loss, it is vital to halt deforestation and implement conservation strategies that protect both forest habitats and the species that depend on them.

Deforestation and Its Economic Consequences for Future Generations

The economic consequences of deforestation are far-reaching and extend well beyond the immediate loss of timber, agricultural land, and resources. While the short-term economic benefits of deforestation may appear substantial, particularly for industries like logging and agriculture, these benefits are unsustainable in the long term. Deforestation can cause the depletion of soil nutrients, which ultimately reduces the productivity of the land and requires costly inputs such as fertilizers. Furthermore, deforestation leads to the loss of ecosystem services such as water regulation, pollination, and carbon sequestration, which have immense economic value. For example, the global economic value of the ecosystem services provided by forests is estimated to be trillions of dollars annually (Costanza et al., 1997). The loss of these services due to deforestation results in long-term economic costs for both local communities and global economies. Additionally, deforestation contributes to climate change, which can increase the frequency and intensity of extreme weather events, further damaging infrastructure, reducing agricultural yields, and increasing the costs of disaster recovery. By adopting sustainable development practices and valuing forest ecosystems, future generations can avoid the economic consequences of deforestation and ensure a more resilient and prosperous future.

Deforestation and Global Policy Solutions

Addressing deforestation on a global scale requires coordinated policy solutions that balance environmental, economic, and social interests. While many countries have made efforts to reduce deforestation through national policies and international agreements, more needs to be done to halt the global trend. One effective approach is the integration of deforestation prevention into broader environmental and climate policies. For instance, international agreements such as the United Nations' REDD+ (Reducing Emissions from Deforestation and Forest Degradation) program incentivize countries to protect their forests by providing financial rewards for forest conservation. REDD+ aims to reduce carbon emissions from deforestation by providing financial support to countries that protect their forests and promote sustainable land-use practices. In addition to international initiatives, local and national policies must address the root causes of deforestation, such as illegal logging, unsustainable agricultural practices, and land tenure issues. Strengthening law enforcement, promoting sustainable land management, and empowering indigenous communities are critical components of any comprehensive deforestation policy. By working together at both global and local levels, governments can implement policies that protect forests, enhance livelihoods, and mitigate climate change.

The Path Forward: Reforestation and Sustainable Land Management

To combat the effects of deforestation and safeguard the environment for future generations, reforestation and sustainable land management practices are essential. Reforestation involves planting trees in areas where forests have been cleared, while afforestation refers to planting trees in areas that were previously not forested. These strategies can help restore carbon sinks, improve biodiversity, and regenerate ecosystems. Additionally, sustainable land management practices, such as agroforestry, can help reduce the pressure on forests by integrating tree planting into agricultural systems, providing both economic benefits and environmental protection. Reforestation not only helps mitigate climate change by absorbing CO2, but it also provides numerous other benefits, including improved water quality, soil stability, and wildlife habitat. Governments, businesses, and local communities must work together to implement reforestation programs, enforce sustainable land management policies, and support forest conservation efforts. Through collective action, it is possible to reverse the damage caused by deforestation and ensure a sustainable future for people and the planet.

Chapter 4: Biodiversity Loss: Importance of species

Introduction to Biodiversity and Species Importance

Biodiversity refers to the variety and variability of life on Earth, encompassing species, ecosystems, and genetic diversity. The loss of biodiversity has profound implications for ecosystems and human societies. Species play critical roles in ecosystem functions such as pollination, nutrient cycling, and climate regulation. When a species is lost, it can disrupt these functions, leading to cascading effects on the environment and human well-being. The importance of species extends beyond their ecological roles; they also provide direct benefits to humans, including medicinal resources, food, and materials. As human activities contribute to habitat destruction, overexploitation, and climate change, the rate of species loss accelerates, threatening global biodiversity. This loss of species not only impacts the natural world but also has significant economic, social, and cultural consequences. Understanding the importance of species is essential for developing strategies to mitigate biodiversity loss and promote conservation efforts.

The Role of Species in Ecosystem Stability

Species contribute to ecosystem stability by maintaining the balance of ecological processes. The presence of a diverse range of species ensures the resilience of ecosystems in the face of environmental stressors. For example, species that act as pollinators, such as bees, are essential for the reproduction of many plants, including crops that humans rely on for food. In addition to pollination, species play roles in pest control, water filtration, and soil fertility. When species are lost, these ecosystem services can decline or collapse, leading to negative consequences for other species, including humans. Biodiversity loss disrupts these complex interactions, making ecosystems more vulnerable to disturbances such as diseases, climate change, and natural disasters. Furthermore, the loss of keystone species—those that have a disproportionately large impact on their environment—can trigger a chain reaction, leading to the decline of other species and the breakdown of ecosystem services. As such, the protection of species and their habitats is critical for maintaining ecosystem functions that support life on Earth.

Economic Value of Species: Ecosystem Services

The economic value of species extends beyond their direct use in agriculture, medicine, and industry. Ecosystem services, which are the benefits that ecosystems provide to humanity, are essential to economic prosperity. These services include clean water, fertile soil, climate regulation, and carbon sequestration, all of which are influenced by the presence of diverse species. For example, the pollination of crops by bees and other insects is a service that directly contributes to food production and the agricultural economy. In the U.S. alone, the economic value of pollination services has been estimated to be worth billions of dollars annually. Similarly, forests, which are home to a vast array of species, provide critical services such as carbon storage and water regulation, contributing to climate change mitigation and the preservation of water resources. The loss of species reduces the capacity of ecosystems to provide these services, resulting in increased costs for human societies in terms of agricultural losses, health impacts, and climate-related disasters. Protecting biodiversity is thus a sound economic strategy, as it helps ensure the continued provision of ecosystem services that are vital for human well-being.

Species and Human Health: Medicinal Benefits

Many species provide medicinal resources that are critical to human health. Natural products derived from plants, animals, and microorganisms have been the foundation of traditional medicine for millennia and continue to play an essential role in modern pharmaceutical development. For instance, the pain-relieving drug morphine is derived from the opium poppy, and the anticancer agent paclitaxel is derived from the bark of the Pacific yew tree. Over half of all prescription drugs used today are based on natural products. Biodiversity loss, particularly the extinction of plant and animal species, threatens the discovery of new medicines that could help combat diseases, including cancer, HIV/AIDS, and malaria. Furthermore, the genetic diversity within species provides valuable resources for developing new treatments and improving public health. As species are lost, the potential for discovering life-saving medicines diminishes, making biodiversity conservation not only an environmental priority but also a health imperative. Efforts to protect and conserve biodiversity are therefore directly linked to improving human health outcomes, highlighting the interconnectedness of species preservation and public well-being.

Climate Regulation: The Role of Species

Species play a vital role in regulating the Earth's climate, particularly through their involvement in carbon sequestration. Forests, wetlands, and oceans are home to a diverse range of species that absorb and store carbon dioxide, helping to mitigate the effects of climate change. Trees, for example, are highly effective at sequestering carbon, with mature forests acting as major carbon sinks. Marine species such as phytoplankton and kelp also contribute to the carbon cycle by absorbing carbon dioxide from the atmosphere and storing it in ocean waters. However, the loss of species can disrupt these processes, leading to a net release of carbon into the atmosphere and exacerbating global warming. For instance, deforestation not only eliminates the trees that capture carbon, but it also reduces the biodiversity that supports forest ecosystems, making them less resilient to environmental stressors. Similarly, the destruction of coastal ecosystems, such as mangroves and seagrasses, undermines their ability to store carbon and protect coastal areas from storm surges and erosion. Protecting and restoring ecosystems rich in biodiversity is therefore crucial for maintaining the Earth's climate stability.

Loss of Species and Ecosystem Services

The loss of species directly threatens the ecosystem services that support life on Earth. Ecosystem services refer to the benefits that humans derive from ecosystems, such as food, clean water, and air purification, which are all influenced by the health and diversity of species. For instance, wetlands, which are rich in biodiversity, filter pollutants from water, regulate floods, and provide habitat for wildlife. When species within these ecosystems are lost, their ability to perform these essential services declines, leading to negative consequences for human health and economic well-being. One example is the decline in bee populations, which has severe implications for crop pollination and, consequently, food production. Similarly, the loss of tree species in tropical rainforests can disrupt the water cycle, resulting in changes in rainfall patterns and soil erosion. Biodiversity loss also makes ecosystems less resilient to environmental changes, such as climate change, because diverse ecosystems are more capable of adapting to stress. Therefore, preserving species is not only essential for ecological balance but also for safeguarding the natural services that humanity depends on for survival.

Species Extinction and Its Impact on the Food Chain

Species extinction disrupts the food chain, which in turn affects the stability of ecosystems. The food chain is a complex web of interactions between organisms, where each species plays a role in maintaining the balance of populations. For example, the loss of predators can lead to the overpopulation of prey species, which can then deplete vegetation and affect the entire ecosystem. Similarly, the loss of herbivores can alter plant communities, leading to changes in the structure of ecosystems. When keystone species—those that have a disproportionately large impact on their environment—are removed, the cascading effects can be devastating. For example, the decline of wolves in Yellowstone National Park led to an increase in elk populations, which caused overgrazing of vegetation and negatively impacted other species in the park. Such imbalances in the food chain highlight the interconnectedness of species and the importance of protecting biodiversity to maintain ecosystem integrity.

The Role of Biodiversity in Economic Development

Biodiversity is not only a pillar of environmental health but also a critical driver of economic development. Healthy ecosystems contribute to economic growth by providing goods and services that are essential for human well-being. For instance, agricultural production relies heavily on biodiversity for soil fertility, water regulation, and pest control. Similarly, the fishing industry depends on the health of aquatic ecosystems, which provide food and livelihoods for millions of people. The tourism industry, especially in areas rich in wildlife and natural landscapes, is another sector that depends on biodiversity. Furthermore, biodiversity provides the raw materials for pharmaceuticals, biofuels, and other industries that contribute to economic growth. The degradation of biodiversity, on the other hand, leads to economic losses due to reduced agricultural yields, damaged infrastructure from flooding, and the loss of income from tourism and fishing. By investing in biodiversity conservation, economies can ensure sustainable development that benefits both people and the environment.

Climate Change and Biodiversity Loss

Climate change is one of the most significant drivers of biodiversity loss in the modern era. Rising temperatures, shifting precipitation patterns, and extreme weather events are altering the habitats of many species, making it increasingly difficult for them to survive in their current environments. For instance, coral reefs, which are highly sensitive to changes in temperature, are experiencing widespread bleaching as a result of rising ocean temperatures. Similarly, many species are being forced to migrate to cooler areas or higher altitudes, but this is not always a viable option, especially for species with limited mobility or those dependent on specific ecosystems. Climate change can also exacerbate the effects of other threats to biodiversity, such as habitat destruction and overexploitation, leading to the accelerated extinction of species. To address biodiversity loss, it is essential to mitigate climate change by reducing greenhouse gas emissions and adopting strategies that help species adapt to changing conditions, such as habitat restoration and the creation of protected areas.

Deforestation and Its Impact on Biodiversity

Deforestation is one of the leading causes of biodiversity loss, particularly in tropical regions where ecosystems are rich in species diversity. Forests provide habitats for countless species, and their destruction leads to the displacement or extinction of these species. The clearing of forests for agriculture, logging, and infrastructure development not only removes habitats but also disrupts ecological processes such as nutrient cycling, carbon sequestration, and water regulation. Deforestation contributes to climate change by releasing stored carbon into the atmosphere, further exacerbating environmental challenges. The Amazon Rainforest, often referred to as the "lungs of the Earth," plays a crucial role in global climate regulation and is home to millions of species, many of which are yet to be discovered (Fearnside, 2005). The loss of such forests undermines the services they provide and accelerates species extinction. Moreover, deforestation fragments ecosystems, isolating populations of species and reducing genetic diversity, which can make them more vulnerable to diseases and environmental changes.

To prevent further biodiversity loss, it is essential to implement sustainable land-use practices, promote reforestation, and protect remaining forests.

Invasive Species and Their Impact on Native Biodiversity

Invasive species are one of the major threats to native biodiversity. These species, which are introduced to new environments—either accidentally or intentionally—can outcompete native species for resources, disrupt food webs, and alter ecosystems. Invasive species often thrive in the absence of natural predators or competitors, allowing them to proliferate rapidly and cause significant ecological damage. For example, the introduction of the brown tree snake to Guam led to the extinction of several bird species on the island, disrupting the local ecosystem. Similarly, the spread of invasive plant species, such as kudzu in the southeastern United States, has resulted in the displacement of native vegetation and the degradation of habitats. Invasive species can also introduce new diseases, further threatening native species. Managing invasive species is essential for conserving biodiversity, and strategies include early detection, rapid response, and habitat restoration to support native species.

Overexploitation of Species and Its Consequences

Overexploitation of species, whether through hunting, fishing, or harvesting, is a significant driver of biodiversity loss. Overfishing, for example, has led to the depletion of fish stocks in many parts of the world, threatening the survival of marine species and disrupting the balance of marine ecosystems (Pauly et al., 2002). Similarly, illegal wildlife trade and hunting have caused the decline of iconic species such as elephants and rhinoceroses, which are targeted for their ivory and horns. Overexploitation not only directly reduces species populations but also weakens ecosystems by removing important members of food webs, which can have ripple effects throughout the environment. For instance, the overhunting of apex predators like tigers can lead to an overabundance of herbivores, which can then overgraze vegetation and reduce plant diversity. Sustainable use of natural resources is essential to prevent overexploitation and protect biodiversity for future generations.

The Role of Protected Areas in Biodiversity Conservation

Protected areas play a crucial role in conserving biodiversity by providing safe havens for species and ecosystems. These areas, such as national parks, wildlife reserves, and marine protected areas, offer habitat protection, mitigate threats from human activities, and contribute to the preservation of biodiversity in the long term. Protected areas are often located in regions of high species diversity, ensuring that vulnerable and endangered species have a chance to survive and recover. For example, the establishment of the Yellowstone National Park in the United States has contributed to the conservation of species such as bison and grizzly bears, which were once on the brink of extinction. Additionally, protected areas help maintain ecosystem services, such as water filtration, carbon sequestration, and soil fertility, which are vital for human well-being. While protected areas alone cannot guarantee the survival of all species, they are a key component of biodiversity conservation strategies and help to ensure the resilience of ecosystems in the face of environmental challenges.

The Importance of Genetic Diversity for Species Survival

Genetic diversity is a critical component of biodiversity that underpins the ability of species to adapt to changing environmental conditions. Populations with high genetic diversity are more likely to survive and thrive in the face of challenges such as disease, climate change, and habitat loss. Genetic diversity enables species to evolve and adapt to new environments over time, increasing their chances of long-term survival. For example, the genetic variation within a population of wild plants or animals allows them to resist pests or diseases that might otherwise decimate a genetically uniform population. On the other hand, populations with low genetic diversity are more susceptible to inbreeding depression, which can lead to a reduction in fitness and an increased risk of extinction. Conservation efforts that focus on maintaining or enhancing genetic diversity, such as creating wildlife corridors and managing breeding programs, are essential for the preservation of species and ecosystems.

Conservation Strategies: In Situ vs Ex Situ Protection

Conservation strategies for protecting species can be broadly categorized into two approaches: in situ and ex situ conservation. In situ conservation involves protecting species in their natural habitats, ensuring that ecosystems remain intact and species can continue to evolve and interact with other species. This approach includes the establishment of protected areas, the restoration of degraded habitats, and the reduction of human impacts on wildlife. On the other hand, ex situ conservation involves the preservation of species outside their natural habitats, such as in zoos, botanical gardens, or seed banks. This approach is often used for species that are at high risk of extinction and cannot survive in the wild due to habitat loss or other threats. While ex situ conservation can provide a safety net for endangered species, in situ conservation is generally preferred because it preserves the full range of ecological interactions and genetic diversity. A combination of both approaches is often necessary to ensure the survival of species and ecosystems.

The Role of Local Communities in Biodiversity Conservation

Local communities play a crucial role in biodiversity conservation, as they often live in close proximity to the ecosystems that are most threatened. Engaging these communities in conservation efforts can lead to more sustainable and effective outcomes. Traditional knowledge and practices that have evolved over centuries can provide valuable insights into how to manage natural resources and protect biodiversity. For example, indigenous communities in the Amazon have long practiced sustainable hunting and fishing techniques that help maintain the balance of species. Moreover, when local communities are involved in conservation initiatives, they are more likely to support and sustain these efforts. In some cases, community-based conservation has proven to be highly successful, such as in the case of community-managed marine protected areas in the Pacific Islands, which have helped restore fish populations. However, for these initiatives to succeed, it is important to provide local communities with the resources, training, and support they need to manage biodiversity sustainably. Empowering local people to take an active role in conservation can help ensure that biodiversity is protected for future generations.

The Impact of Agriculture on Biodiversity

Agriculture has long been a major driver of biodiversity loss. The expansion of agricultural land, particularly for monoculture crops like soy and palm oil, leads to the destruction of natural habitats, forcing species to relocate or face extinction. Pesticides and fertilizers used in industrial agriculture can pollute water sources, harm non-target species, and reduce the diversity of soil organisms. Furthermore, the intensification of livestock farming has contributed to land degradation and deforestation, particularly in tropical regions, where forests are cleared to create grazing land or grow animal feed. Although sustainable agricultural practices, such as agroforestry and organic farming, can help mitigate some of these impacts, the pressure of feeding a growing global population remains a challenge. The integration of biodiversity conservation into agricultural practices is essential to reduce the negative effects of farming on ecosystems and ensure that agriculture can coexist with healthy biodiversity.

The Economic Value of Biodiversity

Biodiversity provides significant economic benefits, not only through the direct exploitation of species but also through the ecosystem services that support agriculture, industry, and human well-being. The total economic value of biodiversity can be difficult to quantify, but it is widely accepted that biodiversity loss could have severe economic consequences. Ecosystem services such as pollination, water purification, and climate regulation contribute trillions of dollars to the global economy every year. The pharmaceutical industry alone depends heavily on the diversity of plants, animals, and microorganisms for the development of medicines. Additionally, industries such as tourism, agriculture, and forestry rely on healthy ecosystems for their sustainability. By investing in biodiversity conservation, governments and businesses can ensure the long-term availability of these ecosystem services, which are essential for economic stability. The growing recognition of the economic value of biodiversity has led to calls for policies that integrate environmental conservation into economic planning and decision-making.

The Role of Education in Biodiversity Conservation

Education plays a crucial role in raising awareness about the importance of biodiversity and the need for its conservation. By educating individuals, communities, and future leaders about the value of biodiversity, we can foster a deeper understanding of the interconnectedness of life on Earth. Environmental education programs in schools, universities, and community organizations can teach people about the impacts of human activities on ecosystems and the species that inhabit them. Public awareness campaigns, eco-tourism initiatives, and citizen science projects are also effective ways to engage the public in biodiversity conservation efforts. Furthermore, education can empower people to take action, whether through reducing their environmental footprint, supporting conservation policies, or participating in local conservation initiatives. By integrating biodiversity conservation into education at all levels, we can create a generation of environmentally conscious individuals who are equipped to tackle the challenges of biodiversity loss.

Concluding Thoughts on Biodiversity Loss and Conservation

Biodiversity loss is one of the most urgent challenges facing humanity today. The depletion of species and ecosystems threatens not only the stability of the environment but also the economic and social well-being of people around the world. The drivers of biodiversity loss—deforestation, climate change, overexploitation, and habitat destruction—are largely human-induced, making it imperative for us to take collective action to address these threats. Conservation strategies, such as protecting natural habitats, restoring degraded ecosystems, and promoting sustainable land-use practices, are essential to halting biodiversity loss. However, these efforts will only be successful if we also address the underlying drivers of biodiversity loss, such as unsustainable consumption and economic development models that prioritize short-term gain over long-term sustainability. By recognizing the intrinsic value of biodiversity and its importance for human survival, we can work together to create a more sustainable future for all species on Earth.

Chapter 5: Renewable Energy: Transitioning to green power

Introduction to Renewable Energy

Renewable energy is energy derived from natural sources that are continuously replenished. Unlike fossil fuels, renewable sources such as solar, wind, and hydropower are abundant and have minimal environmental impacts. The transition from traditional, carbon-intensive energy sources to renewable energy is essential in mitigating climate change and achieving sustainable energy systems. As global energy demand increases, transitioning to renewable energy offers an opportunity to reduce greenhouse gas emissions, improve energy security, and stimulate economic growth. This transition is vital in reducing the reliance on non-renewable energy sources like coal, oil, and natural gas, which not only pollute the environment but also contribute significantly to global warming. The widespread adoption of renewable energy is supported by technological advances, policy incentives, and growing awareness of the environmental and economic benefits of cleaner energy. However, challenges such as energy storage, infrastructure development, and initial capital investments must be addressed to ensure a smooth and efficient transition to green power.

The Importance of Transitioning to Green Power

The transition to renewable energy is critical in addressing the urgent challenges posed by climate change. Fossil fuels, including coal, oil, and natural gas, are responsible for the majority of global greenhouse gas emissions, leading to rising global temperatures, more extreme weather events, and sea-level rise. Renewable energy sources, such as solar, wind, and hydropower, produce little or no emissions, making them key solutions in mitigating climate change. Furthermore, renewable energy offers enhanced energy security by reducing dependence on imported fossil fuels and promoting decentralized energy systems. This decentralization can provide power to rural or remote areas, improving access to electricity and driving economic development in underserved regions. The benefits of transitioning to renewable energy extend beyond environmental considerations; they also include job creation, economic growth, and improved public health outcomes. For instance, the renewable energy sector has been a major driver of new employment opportunities, from manufacturing solar panels to the installation and maintenance of wind turbines. Therefore, transitioning to green power offers an opportunity to reduce environmental harm while promoting sustainable economic growth and social well-being.

Solar Energy: A Growing Solution

Solar energy, one of the most abundant and widely used renewable energy sources, plays a pivotal role in the global energy transition. Solar power harnesses the sun's energy through photovoltaic (PV) cells, which convert sunlight into electricity. The efficiency and cost-effectiveness of solar panels have significantly improved in recent years, making solar power more accessible and competitive with traditional energy sources. Solar energy is scalable, with applications ranging from small rooftop installations to large utility-scale solar farms. One of the key benefits of solar energy is its minimal environmental impact during operation. Unlike fossil fuels, solar energy does not produce harmful emissions or pollutants. However, challenges remain, particularly related to energy storage. Solar power is intermittent, as it depends on sunlight, which is not available 24/7. To address this, advances in battery storage technology, such as lithium-ion and other emerging energy storage systems, are improving the ability to store solar energy for later use. As solar energy becomes more affordable and efficient, it is expected to play an increasingly larger role in the global renewable energy mix.

Wind Energy: Tapping into the Power of the Wind

Wind energy, like solar power, is a clean and abundant source of renewable energy. Wind turbines convert the kinetic energy of wind into mechanical power, which is then used to generate electricity. Wind energy is highly scalable, from small, decentralized turbines to large offshore wind farms. One of the key advantages of wind energy is its low operating costs once the infrastructure is in place. The potential for wind energy is vast, particularly in areas with strong, consistent winds, such as coastal regions and high plains. Offshore wind farms, which are located in bodies of water, have the added benefit of stronger and more predictable winds, making them an increasingly popular option for large-scale wind energy generation. However, challenges such as intermittency, noise concerns, and land use issues remain barriers to wind energy adoption. Despite these challenges, technological advancements in turbine design and energy storage solutions are helping to make wind energy a key component of the global renewable energy transition.

Hydropower: Harnessing the Energy of Water

Hydropower, or hydroelectric power, is one of the oldest and most widely used forms of renewable energy. It generates electricity by using the kinetic energy of flowing or falling water to turn turbines that drive generators. Hydropower accounts for a significant portion of the world's renewable energy capacity, providing reliable and large-scale electricity generation (Zhao et al., 2020). While hydropower is a highly efficient and consistent energy source, it does have environmental and social challenges, especially in the case of large dams. Dams can disrupt local ecosystems, affect water quality, and displace communities. However, smaller-scale hydropower projects and innovations in low-impact hydropower are helping to mitigate some of these challenges. Despite these concerns, hydropower remains a key player in the transition to green power, offering a consistent and scalable source of renewable energy.

Geothermal Energy: Tapping the Earth's Heat

Geothermal energy is derived from the heat stored beneath the Earth's surface. This heat can be used to generate electricity or for direct heating applications. Geothermal power plants are typically located in regions with significant geothermal activity, such as volcanic areas (Lund & Boyd, 2016). One of the key benefits of geothermal energy is its reliability. Unlike solar and wind, geothermal energy is not subject to fluctuations in weather or time of day, providing a consistent and predictable energy source. However, geothermal energy is geographically limited to areas with accessible heat sources, and initial capital costs for drilling and plant construction can be high. Despite these challenges, geothermal energy plays an important role in the transition to green power by providing low-emission, base-load energy that can complement other renewable sources.

Bioenergy: Renewable Power from Organic Materials

Bioenergy refers to the energy derived from organic materials such as plants, agricultural waste, and algae. Biomass can be converted into biofuels, electricity, or heat through combustion, anaerobic digestion, or fermentation processes. One of the most common forms of bioenergy is biofuels, which are used in transportation to reduce the reliance on fossil fuels. Bioethanol and biodiesel are the primary biofuels produced from crops such as corn, sugarcane, and soybeans. The main advantage of bioenergy is that it can be stored and used as a flexible power source, complementing other renewable energies like wind and solar. However, large-scale bioenergy production can lead to concerns such as food security, land use changes, and deforestation. Furthermore, while biofuels are considered renewable, their lifecycle emissions depend on the feedstock and production processes used. Despite these challenges, bioenergy remains a key component in achieving a diversified, low-carbon energy system.

Energy Storage Solutions for Renewable Energy

As renewable energy sources such as solar and wind are intermittent, energy storage technologies are crucial to ensuring a stable and reliable power supply. Energy storage allows excess energy generated during periods of high production to be stored and released when demand is higher or when renewable generation is low. Battery storage technologies, such as lithium-ion batteries, have seen significant improvements in energy density, cost, and lifespan, making them a viable option for large-scale energy storage. Additionally, pumped hydro storage, compressed air energy storage, and thermal storage systems are gaining traction as complementary technologies for renewable energy systems. The development of these energy storage solutions is essential in facilitating the widespread adoption of renewable energy and ensuring grid stability. With ongoing research and innovation, energy storage will continue to evolve, supporting the global transition to clean, renewable power.

The Role of Smart Grids in Renewable Energy Integration

Smart grids are modernized electrical grids that leverage digital communication technologies to monitor and control energy flows in real-time. These grids enable a more efficient and reliable integration of renewable energy sources by balancing supply and demand dynamically. By using smart meters and sensors, smart grids can detect energy consumption patterns and make adjustments to optimize the distribution of renewable power, ensuring that energy is used where it is most needed (Liu et al., 2018). The ability to integrate decentralized energy sources, such as solar and wind, with traditional power plants allows for a more resilient energy infrastructure. Furthermore, smart grids enable demand response, where consumers can adjust their energy usage based on price signals or grid needs, promoting energy conservation. The deployment of smart grid technologies is an essential step in transitioning to a green power system that maximizes the potential of renewable energy sources while maintaining grid reliability and efficiency.

Policy and Regulatory Support for Renewable Energy

Effective policies and regulations are fundamental in driving the transition to renewable energy. Governments play a crucial role in establishing supportive frameworks that encourage investment in clean energy technologies, facilitate innovation, and create incentives for renewable energy adoption. In many countries, renewable energy policies include subsidies, tax incentives, and feed-in tariffs, which guarantee fixed payments for renewable energy producers. Additionally, international agreements such as the Paris Agreement provide a global framework for addressing climate change by reducing carbon emissions and promoting sustainable energy solutions. While significant progress has been made in renewable energy policy, challenges remain, such as overcoming vested interests in fossil fuels and ensuring a just transition for communities and workers impacted by the shift to green energy. Policymakers must balance environmental, economic, and social objectives to foster an inclusive and sustainable energy future.

Transitioning to Green Power: Challenges and Opportunities

While the transition to renewable energy offers immense potential for mitigating climate change and achieving sustainability, it also presents several challenges. One of the most significant hurdles is the need for substantial upfront investments in renewable energy infrastructure, including power plants, transmission networks, and storage systems. Additionally, renewable energy technologies are often intermittent, requiring advanced grid management and energy storage solutions to ensure reliable supply. Furthermore, the transition to green power requires a shift in societal behaviors, including changes in energy consumption patterns and support for clean energy policies. Governments, industry leaders, and consumers must work collaboratively to overcome these challenges and capitalize on the opportunities presented by renewable energy. Addressing these barriers can unlock the full potential of renewable energy to provide a cleaner, more sustainable energy future for all.

The Global Impact of Renewable Energy Adoption

The widespread adoption of renewable energy has far-reaching global impacts, including the potential to reduce greenhouse gas emissions and improve public health by minimizing air pollution. By replacing fossil fuels with clean energy sources, countries can significantly reduce their carbon footprints, helping to meet international climate targets such as those set by the Paris Agreement. Additionally, renewable energy can contribute to energy security by diversifying energy sources and reducing reliance on imported fossil fuels. The shift to green power can also stimulate job creation and economic development, particularly in sectors such as clean technology innovation, renewable energy infrastructure development, and energy efficiency services. The global transition to renewable energy is crucial in addressing climate change while fostering a sustainable and equitable energy future.

The Path Forward for Green Power

The transition to renewable energy is a critical step in addressing climate change and achieving global sustainability. With technological advances, supportive policies, and growing public awareness, renewable energy sources such as solar, wind, hydropower, geothermal, and bioenergy can significantly reduce our reliance on fossil fuels and minimize environmental degradation. While challenges remain, the potential benefits of renewable energy—ranging from reduced emissions to improved energy security—offer compelling reasons to continue the global shift toward green power. By overcoming technological, economic, and societal barriers, we can pave the way for a cleaner, more sustainable future powered by renewable energy.

Solar Power: Leading the Renewable Revolution

Solar power has become one of the most dominant forms of renewable energy due to its abundant availability and rapidly decreasing costs. Solar panels convert sunlight directly into electricity, making solar energy an essential component of the global green energy transition (International Energy Agency [IEA], 2020). The technology behind photovoltaic (PV) cells has improved significantly in recent years, increasing their efficiency and lifespan. As a result, the use of solar power has expanded from small-scale residential installations to large-scale solar farms. Solar energy offers numerous benefits, including reducing dependence on fossil fuels, lowering greenhouse gas emissions, and creating local jobs. However, solar power generation is intermittent, relying on weather conditions and the time of day. As such, it often requires complementary energy storage solutions and integration with other renewable sources to provide a reliable, constant energy supply (Sims et al., 2007). Despite these challenges, solar energy remains a key player in achieving a sustainable, low-carbon energy future.

Wind Power: Harnessing Nature's Strength

Wind power is another prominent renewable energy source that is rapidly gaining traction worldwide. Wind turbines capture the kinetic energy from the wind and convert it into electricity, providing a clean and sustainable source of power (Kumar et al., 2019). One of the main advantages of wind energy is its ability to generate power even in remote locations, making it a viable option for regions with little access to centralized energy grids. In addition, onshore and offshore wind farms are now being developed in various parts of the world, contributing significantly to the diversification of the global energy mix (IEA, 2020). However, wind energy production is also intermittent, as it depends on wind speed and availability. Despite these challenges, advances in turbine design, energy storage, and grid integration are helping to overcome limitations and maximize the potential of wind power (Akorede et al., 2010). Overall, wind energy plays a pivotal role in the transition to green power and a sustainable energy future.

Hydropower: A Reliable Source of Green Power

Hydropower is one of the oldest and most widely used forms of renewable energy. It generates electricity by converting the kinetic energy of flowing water into mechanical energy through turbines (World Energy Council, 2019). Hydropower plants can range from large-scale dams to small, decentralized systems known as run-of-the-river plants. One of the key advantages of hydropower is its reliability; unlike solar and wind power, hydropower can provide a consistent supply of energy, often referred to as "baseload" power (IEA, 2020). However, hydropower has environmental impacts, particularly related to the construction of large dams, which can disrupt ecosystems and displace local communities (WCD, 2000). Moreover, hydropower relies on the availability of water, which may be affected by climate change, reducing its potential in regions facing water scarcity. Nevertheless, with appropriate management, hydropower continues to be a valuable renewable energy source in the global transition to green power.

Geothermal Energy: Tapping into the Earth's Heat

Geothermal energy harnesses the heat stored beneath the Earth's surface to generate electricity or provide direct heating. Geothermal power plants use steam or hot water from underground reservoirs to drive turbines connected to generators (Lund et al., 2010). This form of energy is reliable, sustainable, and produces minimal greenhouse gas emissions compared to fossil fuel-based power generation (Lund & Boyd, 2016). Geothermal energy is particularly advantageous for countries with active volcanic regions, such as Iceland, the

Philippines, and parts of the United States (IEA, 2020). However, geothermal energy is geographically limited, as it requires specific geological conditions to be viable. Additionally, the initial costs of drilling and infrastructure can be high, making it less accessible for certain regions. Despite these challenges, geothermal energy remains an important part of the renewable energy landscape, particularly for countries seeking to reduce their carbon emissions and enhance energy security.

The Role of Nuclear Power in the Renewable Energy Mix

While not traditionally considered a renewable energy source, nuclear power can play an important role in reducing greenhouse gas emissions when integrated into a broader low-carbon energy mix. Nuclear power plants generate electricity through the process of nuclear fission, producing large amounts of energy with minimal carbon emissions (IAEA, 2020). One of the key benefits of nuclear power is its ability to provide baseload power, similar to hydropower, making it a reliable complement to intermittent renewable energy sources such as wind and solar (Sovacool, 2011). However, nuclear energy faces significant challenges, including concerns about safety, radioactive waste disposal, and the high costs associated with plant construction and decommissioning. Despite these challenges, some countries, such as France, have successfully integrated nuclear power into their energy mix as a low-carbon alternative to fossil fuels. The future of nuclear power will depend on technological advancements in reactor design and waste management, as well as public acceptance.

Integrating Renewable Energy into Existing Power Grids

As renewable energy continues to grow, integrating these sources into existing power grids presents both challenges and opportunities. Traditional power grids were designed to accommodate centralized, fossil-fuel-based energy sources, which operate continuously. In contrast, renewable energy sources such as solar and wind are intermittent, requiring advanced grid management technologies (Kwasinski et al., 2013). Smart grid technologies, along with energy storage solutions, are key to ensuring that renewable energy can be integrated seamlessly into the grid, providing consistent and reliable power (Zhao et al., 2020). Grid modernization also includes the deployment of advanced metering infrastructure (AMI), which helps monitor energy consumption and enables demand-side management. These technological innovations are critical for the successful integration of renewable energy into the global power system, facilitating the transition to a green power future.

Conclusion: A Green Future Powered by Renewable Energy

The transition to renewable energy is essential for mitigating climate change and securing a sustainable future. With growing technological innovations and increasing public and governmental support, renewable energy sources such as solar, wind, hydropower, and geothermal have the potential to transform the global energy landscape (International Renewable Energy Agency [IRENA], 2020). While challenges such as intermittency, high initial costs, and infrastructure requirements remain, the benefits of renewable energy—reduced emissions, energy security, job creation, and sustainable development—make it the cornerstone of a low-carbon future. By continuing to invest in renewable energy technologies, policy frameworks, and grid integration, the world can make a successful transition to green power and build a cleaner, more sustainable energy future.

Chapter 6: Sustainable Agriculture: Feeding the world

Introduction to Sustainable Agriculture

Sustainable agriculture refers to farming practices that meet current food needs while ensuring the health of future generations through environmentally responsible and socially equitable practices (Pretty, 2008). As the global population continues to grow, the pressure to increase food production intensifies, presenting challenges to both natural ecosystems and food security. Unlike conventional farming methods, which often rely on intensive inputs of water, energy, and chemicals, sustainable agriculture focuses on long-term agricultural productivity without compromising the environment (FAO, 2014). This approach involves optimizing land use, minimizing waste, protecting biodiversity, and reducing dependency on non-renewable resources. In this context, sustainable agriculture serves not only as a solution to feed the world's growing population but also as a critical mechanism for achieving environmental conservation and climate resilience (Tilman et al., 2011).

The Need for Sustainable Agricultural Practices

The need for sustainable agricultural practices has never been more urgent, as conventional farming practices often result in soil degradation, water scarcity, and loss of biodiversity (Garnett, 2014). Unsustainable agricultural practices such as monocropping and excessive use of chemical fertilizers and pesticides have led to the depletion of natural resources, such as fertile soil and clean water, which are critical for food production (Foley et al., 2011). Furthermore, the environmental consequences of these practices have contributed to climate change, further exacerbating the challenges faced by the agricultural sector. Sustainable agriculture seeks to reverse these trends by focusing on practices that enhance soil health, conserve water, and promote biodiversity. These practices include crop rotation, agroforestry, and integrated pest management, all of which aim to build a resilient agricultural system that is capable of maintaining food production while minimizing environmental impact (Garnett, 2014).

The Role of Agroecology in Sustainable Agriculture

Agroecology, a scientific discipline that applies ecological principles to agricultural systems, has gained significant attention as an approach to sustainable agriculture. Agroecological practices focus on working with nature to optimize biodiversity and ecosystem services while promoting sustainable food production. For example, agroecology encourages the use of diverse crops to enhance resilience, reduces dependence on external chemical inputs, and emphasizes soil conservation and water management (Gliessman, 2016). By adopting agroecological practices, farmers can increase productivity while mitigating environmental damage, such as soil erosion, water contamination, and greenhouse gas emissions. In addition, agroecology supports food sovereignty by empowering local farmers and communities to manage their land and resources in a sustainable way. This approach has been shown to increase food security, promote economic resilience, and protect the environment.

The Importance of Biodiversity in Sustainable Agriculture

Biodiversity plays a critical role in sustainable agriculture, as diverse ecosystems are essential for maintaining soil fertility, pest control, and pollination services. The loss of biodiversity, particularly in agricultural landscapes, can lead to reduced ecosystem resilience and greater vulnerability to pests, diseases, and climate change. Incorporating biodiversity into agricultural systems, such as through agroforestry, polyculture, and the conservation of natural habitats, can significantly enhance farm productivity and environmental sustainability. Diverse agricultural systems are also more adaptable to changing climate conditions, providing a buffer against extreme weather events, such as droughts and floods. Furthermore, healthy ecosystems contribute to carbon sequestration, reducing greenhouse gas emissions associated with agriculture. Thus, preserving biodiversity within agricultural landscapes is vital for achieving sustainable and resilient food systems.

Water Management in Sustainable Agriculture

Water is a crucial resource for agriculture, and efficient water management is essential to sustainable farming practices (Postel, 2000). Agriculture accounts for approximately 70% of global freshwater use, and inefficient irrigation practices often lead to water wastage, soil salinization, and reduced crop yields (Foley et al., 2011). Sustainable agricultural practices focus on optimizing water use through techniques such as drip irrigation, rainwater harvesting, and the use of drought-resistant crop varieties (Kijne et al., 2003). These methods not only conserve water but also improve the resilience of crops to water stress, enhancing food security in water-scarce regions. Additionally, sustainable water management practices promote the protection of freshwater ecosystems and improve the long-term viability of agricultural production (Postel, 2000).

Soil Health and Sustainable Farming

Soil health is foundational to sustainable agriculture, as fertile, well-managed soil supports plant growth, maintains water retention, and stores carbon (Lal, 2015). Traditional farming practices, such as excessive tilling and monocropping, can degrade soil structure, reduce organic matter, and deplete essential nutrients, resulting in diminished crop yields and increased vulnerability to environmental stresses (Gattinger et al., 2012). In contrast, sustainable farming practices focus on soil conservation techniques, including crop rotation, cover cropping, reduced tillage, and the use of organic fertilizers (Lal, 2015). These practices help rebuild soil organic matter, increase soil biodiversity, and enhance soil structure, ultimately leading to more resilient and productive farming systems (Gattinger et al., 2012). Healthy soils also contribute to climate change mitigation by sequestering carbon from the atmosphere, reducing the overall carbon footprint of agriculture.

The Role of Technology in Sustainable Agriculture

Advancements in technology have revolutionized the agricultural sector, enabling farmers to adopt more efficient and sustainable practices (Foley et al., 2011). Precision farming, which involves the use of GPS, sensors, and data analytics to optimize farming operations, allows farmers to monitor crop health, soil conditions, and water usage in real time. This technology helps reduce input costs, minimize environmental impact, and improve yields by applying resources only where and when they are needed (Zhang et al., 2002). Moreover, biotechnology has facilitated the development of genetically modified crops that are resistant to pests, diseases, and adverse weather conditions, which can contribute to more sustainable food production in challenging environments (James, 2016). However, the adoption of technology in agriculture must be carefully managed to ensure equitable access, prevent environmental harm, and protect local biodiversity.

Agroforestry: A Key Component of Sustainable Agriculture

Agroforestry, the integration of trees and shrubs into agricultural systems, plays a significant role in promoting sustainability within agriculture. By combining forestry with crop and livestock production, agroforestry systems provide multiple ecological and economic benefits (Jose, 2009). These systems can enhance soil fertility, conserve water, and increase biodiversity by creating habitats for wildlife (Milder et al., 2011). Agroforestry practices, such as alley cropping, silvopasture, and forest farming, help mitigate climate change by sequestering carbon in trees and soils. Additionally, agroforestry improves the resilience of farming systems to climatic extremes, such as droughts and floods, by reducing soil erosion, increasing water infiltration, and protecting crops from wind damage (Jose, 2009). These practices contribute to long-term food security while preserving natural resources and enhancing ecosystem services.

The Role of Government Policies in Promoting Sustainable Agriculture

Government policies play an essential role in fostering sustainable agriculture by incentivizing practices that promote environmental conservation, food security, and rural development (Beddington et al., 2012). Policies such as subsidies for organic farming, crop insurance for sustainable practices, and funding for research on climate-resilient crops can encourage farmers to adopt more sustainable methods (Pretty, 2008). Additionally, land use regulations, such as promoting conservation easements and establishing protected areas for biodiversity, can help safeguard natural resources (FAO, 2014). However, it is crucial that these policies are designed to support smallholder farmers, who often face significant barriers to accessing resources and markets. By integrating sustainability into agricultural policies, governments can create a more resilient and equitable food system that benefits both people and the planet.

The Role of Education and Awareness in Sustainable Agriculture

Education and awareness play pivotal roles in advancing sustainable agricultural practices, particularly in promoting knowledge about the importance of soil health, biodiversity, and climate-resilient farming techniques. Training programs, extension services, and farmer-to-farmer knowledge-sharing initiatives are essential in ensuring that farmers adopt sustainable practices

(Pimbert, 2010). Furthermore, increasing public awareness about the environmental impacts of conventional farming and the benefits of sustainable agriculture can foster consumer demand for sustainably produced food. Educated consumers are more likely to choose food produced using methods that protect the environment, thereby supporting the transition to more sustainable food systems (Horrigan et al., 2002). Ultimately, education empowers both farmers and consumers to make informed decisions that contribute to food security and environmental sustainability.

The Challenges of Scaling Sustainable Agriculture

While sustainable agriculture offers significant benefits, scaling these practices to meet the food demands of a growing global population presents considerable challenges. These challenges include limited access to resources, inadequate infrastructure, and lack of financial support for farmers to transition from conventional to sustainable farming practices (Pretty, 2008). In many parts of the world, farmers lack access to credit, which makes it difficult for them to invest in sustainable technologies or inputs (FAO, 2014). Additionally, the global market often prioritizes short-term productivity over long-term sustainability, which can make it difficult for smallholder farmers to remain competitive. Scaling sustainable agriculture requires significant investments in infrastructure, technology, and education to enable farmers to adopt and maintain sustainable practices while ensuring food security (Tilman et al., 2011).

The Impact of Climate Change on Sustainable Agriculture

Climate change poses a significant threat to the future of sustainable agriculture, as changing weather patterns, such as rising temperatures, altered rainfall patterns, and increased frequency of extreme weather events, can adversely affect crop yields and agricultural productivity (Foley et al., 2011). Sustainable farming practices, such as crop diversification, agroforestry, and water-efficient irrigation, can help mitigate the effects of climate change by enhancing the resilience of agricultural systems (Lipper et al., 2014). However, the unpredictable nature of climate change requires adaptive management strategies that allow farmers to respond to shifting climatic conditions in real time. For example, farmers may need to adjust planting schedules, adopt drought-resistant crop varieties, or invest in soil conservation practices to maintain yields under changing environmental conditions (Foley et al., 2011).

The Role of Organic Farming in Sustainable Agriculture

Organic farming is a cornerstone of sustainable agriculture, as it prioritizes soil health, biodiversity, and ecosystem balance over the use of synthetic chemicals and fertilizers (Reganold & Wachter, 2016). Organic practices such as composting, crop rotation, and the use of natural pesticides and fertilizers reduce the environmental impact of farming by minimizing water and soil contamination. Furthermore, organic farming systems are often more resilient to environmental stresses, such as droughts and pests, due to their emphasis on biodiversity and soil health (Reganold & Wachter, 2016). While organic farming can be labor-intensive and yield lower quantities compared to conventional farming, it offers important benefits, including improved soil quality, reduced environmental degradation, and increased market demand for organic products. As consumer interest in organic products continues to grow, the expansion of organic farming represents a significant opportunity for sustainable food production (Lynch, 2016).

Integrating Sustainability with Global Food Systems

Integrating sustainability with global food systems is essential for addressing both the environmental and social challenges associated with food production. Current global food systems are highly dependent on industrial farming practices, which often lead to environmental degradation, poor working conditions, and inequality in food access (Beddington et al., 2012). Sustainable agriculture seeks to create food systems that are both environmentally sound and socially equitable, ensuring that food production meets the needs of present and future generations. This involves fostering local food systems, supporting smallholder farmers, and promoting fair trade practices. By transforming global food systems to prioritize sustainability, the world can move towards a more just and resilient food future (FAO, 2014).

Supporting Smallholder Farmers in Sustainable Agriculture

Smallholder farmers represent the backbone of food production in many developing countries, producing a significant portion of the world's food, yet they often face immense challenges such as limited access to resources, technology, and markets (FAO, 2014). Supporting smallholder farmers in adopting sustainable agricultural practices is critical for achieving global food security and environmental sustainability. By providing access to financial resources, agricultural extension services, and climate-resilient technologies, governments and organizations can empower smallholder farmers to transition to sustainable farming methods (Pretty, 2008). In addition, cooperatives and farmer groups can help smallholders gain access to larger markets and share knowledge about sustainable practices, thus improving their livelihoods and increasing food security (Chambers & Conway, 1992). Investing in smallholder agriculture is essential for reducing poverty, improving food security, and promoting the sustainable use of natural resources.

The Economic Benefits of Sustainable Agriculture

Sustainable agriculture not only provides environmental and social benefits but also offers significant economic advantages. By reducing input costs, improving soil health, and enhancing water efficiency, sustainable farming practices can increase long-term profitability for farmers (FAO, 2014). Practices such as crop diversification and agroforestry can improve resilience to market fluctuations and climate variability, providing a more stable source of income for farmers (Garnett, 2014). Furthermore, sustainable agriculture promotes the creation of green jobs and supports local economies by reducing dependency on imported chemicals and fossil fuels. In the long run, investing in sustainable agriculture contributes to economic stability by creating more efficient, resilient, and competitive food systems (Pretty, 2008). As consumer demand for sustainably produced food grows, there is also an opportunity for farmers to access premium markets, increasing their profitability and contributing to sustainable development.

Sustainable Agriculture and Food Security

Sustainable agriculture plays a vital role in ensuring long-term food security by producing healthy, nutritious food while protecting the environment and improving farmers' livelihoods (Beddington et al., 2012). By focusing on environmentally friendly practices, such as reducing chemical inputs, improving soil health, and enhancing water conservation, sustainable agriculture helps secure the resources necessary to support future food production. Additionally, sustainable farming practices can increase resilience to climate change, reducing vulnerability to droughts, floods, and other extreme weather events that threaten food systems (Foley et al., 2011). Ultimately, adopting sustainable agriculture can help address the challenges of feeding the growing global population, ensuring food security for future generations.

Scaling Up Sustainable Agriculture Globally

While sustainable agriculture offers numerous benefits, scaling these practices globally remains a significant challenge. Key barriers to scaling include limited access to financial resources, knowledge gaps, and policy constraints that prioritize industrial agriculture (FAO, 2014). However, by creating a supportive policy environment, increasing investment in agricultural research and development, and promoting international cooperation, the global community can work together to scale up sustainable agriculture. Furthermore, promoting the sharing of best practices and lessons learned across regions can help farmers in different parts of the world adopt and adapt sustainable practices to local conditions (Pretty, 2008). As demand for sustainable food increases, businesses, governments, and international organizations have an opportunity to support the transition to a more sustainable food system that benefits people, the planet, and future generations.

The Importance of Local and Regional Food Systems

Local and regional food systems are a critical component of sustainable agriculture, as they promote the consumption of locally grown food, reduce food miles, and support local economies (Hinrichs, 2000). By focusing on small-scale, sustainable farming, local food systems help preserve biodiversity, reduce reliance on industrial agriculture, and provide access to fresh, nutritious food. Additionally, local food systems can enhance food security by reducing dependency on global supply chains that are vulnerable to disruptions caused by climate change or economic instability (Garnett, 2014). Strengthening local and regional food systems also promotes social equity by creating opportunities for smallholder farmers and empowering local communities to manage their food production systems sustainably.

Conclusion: A Future with Sustainable Agriculture

As the world faces growing challenges related to food security, climate change, and environmental degradation, sustainable agriculture offers a path forward that prioritizes long-term environmental, social, and economic health (Foley et al., 2011). By adopting practices such as agroecology, organic farming, and water conservation, we can create food systems that meet the needs of a growing population while protecting the planet's resources. Governments, businesses, and consumers all play important roles in supporting sustainable agriculture through policies, investments, and conscious consumption. The future of food security and environmental sustainability hinges on our ability to transition to sustainable agricultural practices that nourish people, support biodiversity, and ensure a healthy planet for future generations.

Sustainable Agriculture and meeting the growing demand for food

Sustainable agriculture is an approach to farming that seeks to meet the food demands of the present without compromising the ability of future generations to meet their needs. It involves practices that are environmentally sound, economically viable, and socially responsible. The importance of sustainable agriculture has grown in recent decades due to challenges such as climate change, resource depletion, and population growth (Tilman et al., 2011). As the global population is expected to reach nearly 10 billion by 2050, increasing food production while minimizing environmental damage has become a critical concern. Sustainable farming practices, such as crop diversification, agroecology, and reduced chemical inputs, can enhance soil fertility, conserve water, and reduce the carbon footprint of agriculture (Foley et al., 2011). Additionally, sustainable agriculture contributes to biodiversity conservation, soil health, and the resilience of farming systems to climate variability. The challenge ahead is to scale up these practices to meet the growing demand for food without causing further environmental degradation.

The Need for Sustainable Agriculture

The increasing global demand for food, coupled with environmental degradation and climate change, necessitates a shift towards sustainable agricultural practices. Traditional farming methods, which rely heavily on synthetic fertilizers, pesticides, and monocropping, have contributed to soil depletion, loss of biodiversity, and pollution of water sources (Pretty, 2008). Sustainable agriculture seeks to reduce the environmental footprint of food production by using practices that restore and preserve natural resources. These practices include crop rotation, agroforestry, and integrated pest management, which help maintain healthy soils, reduce dependency on chemical inputs, and promote ecological balance. Moreover, sustainable agriculture enhances food security by increasing resilience to climate change and ensuring that farming practices are adaptable to future environmental challenges (Garnett, 2014). The long-term viability of agricultural systems depends on their ability to maintain ecological balance while ensuring equitable access to food for all.

Principles of Sustainable Agriculture

Sustainable agriculture is grounded in several core principles that guide its practices. First, it emphasizes the importance of conserving and enhancing the natural resource base, such as soil, water, and biodiversity. This can be achieved through techniques like agroecology, which integrates ecological principles into farming practices to create more resilient farming systems. Second, sustainable agriculture prioritizes the efficient use of resources, minimizing waste and pollution while optimizing production. Third, it aims to support rural communities by improving livelihoods, providing fair wages, and promoting social equity (FAO, 2014). Additionally, sustainable agriculture fosters local food systems by promoting locally grown, nutritious food and supporting smallholder farmers who are often the most vulnerable to environmental degradation. These principles collectively aim to achieve a balanced approach to food production that meets both the needs of the environment and society.

Climate Change and Sustainable Agriculture

Climate change poses significant challenges to global food security, making sustainable agriculture even more critical. Rising temperatures, changes in precipitation patterns, and extreme weather events such as droughts and floods can directly affect crop yields, soil health, and water availability (Foley et al., 2011). In response to these challenges, sustainable farming practices can help mitigate the effects of climate change by improving soil carbon sequestration, conserving water, and enhancing the resilience of farming systems. Practices such as conservation tillage, agroforestry, and crop diversification can help farmers adapt to changing climatic conditions and maintain food production (Lipper et al., 2014). Moreover, sustainable agriculture can contribute to climate change mitigation by reducing greenhouse gas emissions associated with conventional farming practices, such as the use of synthetic fertilizers and fossil fuel-based machinery (Tilman et al., 2011). Addressing the challenges of climate change requires a shift towards farming systems that are more resilient, adaptable, and capable of sustainably feeding the growing global population.

Soil Health and Sustainable Farming

Soil health is a cornerstone of sustainable agriculture, as healthy soils are essential for producing nutrient-dense food and maintaining ecosystem services. Soil degradation, which results from practices such as monocropping, excessive tillage, and the overuse of chemical fertilizers, threatens both food security and the environment (Pretty, 2008). Sustainable agriculture emphasizes soil conservation practices that improve soil structure, fertility, and biodiversity. These practices include reduced tillage, crop rotation, cover cropping, and organic farming. These methods help build organic matter in the soil, which improves its water-holding capacity, reduces erosion, and supports a diverse range of beneficial organisms (Reganold & Wachter, 2016). By promoting soil health, sustainable agriculture can improve crop yields, reduce input costs, and enhance the resilience of farming systems to climate change.

Water Conservation and Sustainable Agriculture

Water is a critical resource for agriculture, and its efficient use is essential for ensuring the sustainability of food systems. In many regions, water scarcity poses a significant challenge to food production, particularly in areas where irrigation is heavily relied upon (FAO, 2014). Sustainable agriculture aims to improve water use efficiency by promoting practices that conserve water and reduce dependence on irrigation. Techniques such as drip irrigation, rainwater harvesting, and soil moisture management can significantly reduce water waste while maintaining crop productivity (Perry, 2012). Additionally, agroecological practices, such as mulching and cover cropping, help retain soil moisture and prevent evaporation, thereby improving water retention in the soil. By conserving water and improving irrigation efficiency, sustainable agriculture can ensure that water resources are used in a manner that supports long-term food security.

Biodiversity and Sustainable Agriculture

Biodiversity plays a critical role in the resilience of agricultural systems, as diverse ecosystems support pollination, pest control, and nutrient cycling, all of which are essential for healthy food production. However, conventional farming practices that prioritize monocropping and the use of synthetic pesticides and fertilizers have contributed to the loss of biodiversity and the degradation of natural ecosystems. Sustainable agriculture, on the other hand, aims to enhance biodiversity by promoting practices such as crop diversification, agroforestry, and organic farming. These practices help create habitats for beneficial insects, birds, and other wildlife, which contribute to ecosystem services that support food production (Garnett, 2014). By protecting and enhancing biodiversity, sustainable agriculture can improve ecosystem resilience and reduce the need for chemical inputs, making farming systems more environmentally sustainable.

Agroecology as a Sustainable Farming Practice

Agroecology is an integrated approach to farming that applies ecological principles to agricultural systems. This practice emphasizes biodiversity, soil health, and the recycling of nutrients within farming ecosystems. Agroecology aims to create agricultural systems that are more resilient to environmental stressors, such as climate change, and that reduce the dependence on external inputs like synthetic fertilizers and pesticides. Through crop diversification, agroforestry, and reduced tillage, agroecology enhances ecosystem services like pollination and natural pest control (Gliessman, 2015). Furthermore, agroecology promotes the use of local knowledge and practices, empowering farmers to make decisions that are best suited to their specific environmental and socio-economic conditions. This approach not only benefits the environment but also helps to improve food security and farmer livelihoods by fostering more sustainable and resilient food systems.

Organic Farming and Its Role in Sustainability

Organic farming is a key component of sustainable agriculture. This approach to farming avoids synthetic chemicals, such as pesticides and fertilizers, and focuses on building healthy soils, conserving water, and enhancing biodiversity (Reganold & Wachter, 2016). Organic practices include crop rotation, green manure, and the use of organic inputs like compost and manure to improve soil fertility. One of the primary benefits of organic farming is its potential to reduce environmental pollution from chemical inputs, which can contaminate water sources and harm wildlife (Garnett, 2014). Organic farming also offers benefits for human health by reducing exposure to harmful chemicals in food. While yields may be lower in the short term compared to conventional farming, organic farming can result in more resilient farming systems and higher profitability in the long term through premium pricing and market access (Reganold & Wachter, 2016). Organic farming also supports rural communities by promoting local food systems and enhancing food security.

The Role of Technology in Sustainable Agriculture

Technological innovations have the potential to significantly enhance the sustainability of agricultural systems. Precision agriculture, for example, uses technology to optimize the use of inputs like water, fertilizers, and pesticides, reducing waste and minimizing environmental impacts (Gebbers & Adamchuk, 2010). Technologies such as GPS-guided tractors, soil sensors, and drone monitoring can help farmers make more informed decisions, improving efficiency and sustainability. Furthermore, advances in biotechnology, such as genetically modified crops with improved pest resistance or drought tolerance, can help increase yields while reducing the need for chemical inputs (Fedoroff, 2015). While there is some controversy surrounding the use of biotechnology, particularly GMOs, there is consensus that technology can play a vital role in improving the efficiency and sustainability of food production. However, it is important that these technologies are accessible to all farmers, especially smallholders, to ensure equitable and sustainable food systems.

Economic Viability of Sustainable Agriculture

For sustainable agriculture to succeed, it must also be economically viable. Farmers need to adopt practices that not only improve environmental sustainability but also make economic sense in the long run (Pretty, 2008). Sustainable farming practices, such as reduced input costs, lower energy consumption, and enhanced soil fertility, can lead to increased profitability by reducing dependency on expensive chemical inputs and improving long-term yields (Garnett, 2014). Additionally, sustainable agriculture can open new market opportunities for farmers, as consumers increasingly demand food that is produced sustainably. Certification schemes such as Fair Trade and Organic can help farmers access premium markets, providing economic incentives for sustainable farming practices (Hughes, 2009). Furthermore, the economic benefits of sustainable agriculture extend beyond the farm gate, as it can create jobs in rural areas, support local economies, and contribute to food security by ensuring the stability of local food systems.

Addressing Global Hunger with Sustainable Agriculture

Global hunger remains one of the most pressing challenges of the 21st century, with nearly 800 million people suffering from malnutrition (FAO, 2014). Sustainable agriculture has the potential to address this issue by providing a more reliable and equitable food supply. By improving agricultural practices, particularly in developing countries, sustainable farming can increase food production while minimizing the negative impacts on the environment. Practices such as agroecology, organic farming, and integrated pest management can increase yields while reducing dependency on costly and environmentally harmful inputs. Furthermore, sustainable agriculture can enhance food security by fostering more resilient farming systems that are less vulnerable to climate change and market fluctuations (Lipper et al., 2014). Sustainable agriculture provides a pathway to ending hunger by creating more efficient, equitable, and resilient food systems that can feed the world in the long term.

Challenges in Scaling Up Sustainable Agriculture

While sustainable agriculture holds great promise, there are several challenges to scaling up these practices globally. One of the key barriers is the lack of financial resources and access to credit for smallholder farmers, who often lack the capital to invest in sustainable farming techniques (FAO, 2014). Furthermore, the dominance of industrial agriculture, which is heavily subsidized in many countries, creates a competitive disadvantage for farmers who seek to adopt more sustainable practices (Garnett, 2014). There is also a need for greater political will to support sustainable agriculture through policy reforms, subsidies, and incentives that favor sustainable practices. Overcoming these challenges requires coordinated action at the local, national, and international levels, involving farmers, governments, NGOs, and the private sector to create an enabling environment for sustainable agriculture.

The Role of Policy in Promoting Sustainable Agriculture

Policies play a critical role in promoting sustainable agriculture by creating incentives and regulations that support environmentally friendly practices. Governments can promote sustainable farming by providing subsidies or tax incentives for practices that improve soil health, conserve water, and reduce carbon emissions (Pretty, 2008). Additionally, policies that encourage the use of organic fertilizers, agroecology, and agroforestry can help farmers transition from conventional farming systems that rely on harmful chemicals to more sustainable methods (Garnett, 2014). Policymakers also need to invest in research and development to improve sustainable farming techniques, as well as infrastructure to support rural communities and facilitate access to markets. However, policy changes must also be supported by education and training programs to ensure that farmers have the knowledge and resources they need to implement sustainable practices effectively.

The Role of Education in Promoting Sustainable Agriculture

Education is crucial for the widespread adoption of sustainable agricultural practices. Farmers, particularly those in developing countries, need access to training and resources that allow them to understand the benefits and methods of sustainable agriculture. Extension services play a key role in educating farmers, providing them with information about crop rotation, organic farming techniques, and integrated pest management (FAO, 2014). In addition to training farmers, educational initiatives should target policymakers, agricultural scientists, and the public to raise awareness about the importance of sustainability in food production (Garnett, 2014). Universities and research institutions also have a vital role to play in generating new knowledge about sustainable farming techniques and disseminating this information to the broader agricultural community. Public education campaigns can further help shift societal attitudes toward more sustainable consumption and support for environmentally friendly agricultural policies.

The Global Impact of Sustainable Agriculture on Food Security

Sustainable agriculture has the potential to transform global food security by ensuring that food production is not only environmentally sustainable but also socially equitable. By focusing on practices that maintain soil health, conserve water, and enhance biodiversity, sustainable farming can increase food production while mitigating the negative effects of climate change (Lipper et al., 2014). Additionally, sustainable agriculture can contribute to food security by fostering more resilient agricultural systems that are better able to withstand shocks such as droughts, floods, and price volatility. Through sustainable practices, smallholder farmers, who produce much of the world's food, can improve their yields and income, reducing poverty and hunger in rural communities. Ultimately, sustainable agriculture is an essential component of global food security, as it promotes the long-term stability of food systems.

Innovations in Sustainable Farming Practices

The field of sustainable agriculture has witnessed numerous innovations in recent years, driven by advancements in technology and research. One promising area is the development of climate-smart agriculture, which incorporates practices designed to increase productivity while reducing greenhouse gas emissions (Lipper et al., 2014). For instance, conservation tillage, which minimizes soil disturbance, helps sequester carbon in the soil and reduces erosion. Another innovation is the use of drought-resistant crop varieties, which can help mitigate the impact of climate change on water-scarce regions (Fedoroff, 2015). Additionally, the integration of livestock and crops in systems such as agroforestry can help improve biodiversity, reduce pests, and improve nutrient cycling. These innovations offer exciting possibilities for transforming agricultural systems to become more sustainable and resilient in the face of climate change.

The Importance of Local Food Systems in Sustainability

Local food systems are a critical component of sustainable agriculture because they reduce the environmental impact associated with the transportation and distribution of food (FAO, 2014). By growing food locally, farmers can reduce the carbon footprint of their operations while increasing the resilience of food systems to global market fluctuations. Local food systems also strengthen communities by fostering connections between consumers and producers, enhancing food security and supporting local economies. Furthermore, local food systems often prioritize sustainable practices, such as organic farming and the use of renewable resources, as farmers are more likely to adopt practices that preserve their local environment. Supporting local food systems is essential for building sustainable and equitable food systems that benefit both people and the planet.

The Future of Sustainable Agriculture

The future of sustainable agriculture hinges on innovation, education, and policy reform. As the global population continues to grow and climate change poses increasing challenges to food production, the need for sustainable farming practices will become even more urgent (Tilman et al., 2011). Advances in technology, such as precision farming, biotechnologies, and climate-smart agriculture, will be crucial in enhancing the efficiency and sustainability of farming systems (Foley et al., 2011). However, these innovations must be accessible to all farmers, particularly smallholders, to ensure that the benefits of sustainable agriculture are realized globally. Moreover, governments must implement policies that incentivize sustainable practices, protect natural resources, and ensure food security for future generations. By prioritizing sustainability in agricultural systems, we can create a future where food production is both environmentally sound and socially just.

Conclusion

In conclusion, sustainable agriculture offers a viable pathway to feed the world while safeguarding the environment for future generations. By adopting practices that promote soil health, conserve water, enhance biodiversity, and reduce the environmental footprint of food production, we can build a more resilient and equitable global food system. While there are significant challenges in scaling up sustainable agriculture, including financial constraints, policy barriers, and knowledge gaps, the potential benefits are immense. With continued innovation, education, and supportive policies, sustainable agriculture can play a central role in ensuring food security, mitigating climate change, and supporting rural communities worldwide. As the world faces unprecedented challenges in feeding its growing population, sustainable agriculture must be prioritized to ensure that we can nourish the planet without depleting it.

Chapter 7: Water Conservation: Preserving our resources

Introduction to Water Conservation

Water is one of the most vital resources on Earth, essential for sustaining all forms of life. However, the world faces increasing challenges related to water scarcity, pollution, and inefficient use. With the global population continuing to rise, it is crucial to focus on water conservation as a means of ensuring long-term sustainability. Water conservation refers to the practice of managing water resources wisely to avoid waste and preserve them for future generations. It involves efforts to reduce water usage, improve water quality, and promote responsible practices at all levels, from individual households to large-scale industrial operations (UN Water, 2018). This practice is not only essential for addressing the challenges of water scarcity but also plays a key role in preserving ecosystems, ensuring food security, and mitigating the effects of climate change. In this context, water conservation is not just a necessity for future generations but an urgent global responsibility.

The Importance of Water Conservation

Water is a finite resource, and despite its abundance on Earth, less than 1% of it is available for human use as freshwater (Vörösmarty et al., 2010). This highlights the urgent need for water conservation practices. Water is essential for drinking, agriculture, industry, and energy production, yet demand continues to rise, exacerbating the risk of depletion. Climate change is further compounding the issue, leading to shifts in precipitation patterns and more frequent droughts. Water conservation plays a critical role in addressing these challenges by reducing overall demand, improving water quality, and maintaining ecosystems that rely on freshwater sources. Furthermore, water conservation can reduce energy consumption associated with water treatment and distribution, promoting a more sustainable approach to managing resources. By conserving water, we can protect both the environment and the future of human development.

Water Scarcity and its Global Impact

Water scarcity is becoming one of the most critical issues facing the planet today. According to the United Nations, approximately 2.2 billion people globally lack access to safely managed drinking water services (UNICEF & WHO, 2019). Factors such as population growth, urbanization, and industrial expansion are putting increasing pressure on water resources. In many regions, especially in arid and semi-arid zones, water scarcity is a daily reality, with over-extraction and contamination of water sources exacerbating the problem (Falkenmark & Rockström, 2013). Furthermore, the impacts of climate change, such as rising temperatures and irregular rainfall patterns, contribute to the unpredictability of water availability. This makes it even more critical to implement water conservation strategies at all levels, from local communities to national policies, to ensure equitable access to this vital resource. Sustainable water management practices, including water-use efficiency, rainwater harvesting, and ecosystem restoration, are key to addressing water scarcity.

Water Conservation at the Household Level

One of the most effective ways to conserve water is by adopting water-saving practices at the household level. Simple actions, such as fixing leaks, installing water-efficient fixtures, and using water-saving appliances, can significantly reduce water consumption. For example, low-flow faucets and showerheads can reduce water use by as much as 50% without sacrificing comfort (EPA, 2021). Additionally, using water-efficient irrigation systems and practicing rainwater harvesting can further help reduce reliance on municipal water supplies. Education and awareness are critical in encouraging individuals to adopt these practices. Communities can also support water conservation by promoting water-saving habits, such as turning off the tap while brushing teeth or washing dishes. Small changes in everyday behavior can collectively lead to substantial water savings, which are crucial for managing global water resources.

Agricultural Water Use and Conservation Techniques

Agriculture accounts for approximately 70% of global freshwater use, making it the largest sector for water consumption (FAO, 2017). Efficient water management in agriculture is critical for addressing global water scarcity. Techniques such as drip irrigation, which delivers water directly to the roots of plants, and the use of soil moisture sensors can greatly reduce water waste (Postel, 2015). In addition, practices like crop rotation, mulching, and the use of drought-resistant crops can improve water use efficiency and help farmers adapt to water shortages. Integrating water-saving technologies into agricultural practices is essential for meeting the growing food demands of a rapidly increasing global population, while also conserving valuable water resources. Sustainable agricultural practices that optimize water use can help secure food production while minimizing environmental impacts.

Industrial Water Conservation Strategies

Industrial activities are another significant contributor to water consumption. Water is essential in many industrial processes, including cooling, cleaning, and manufacturing. As industries grow, they place increasing pressure on local water resources, leading to conflicts with communities and ecosystems (Gleick, 2014). To address this, industries are adopting water conservation strategies that focus on water reuse and recycling, as well as reducing water consumption in production processes. Technologies such as closed-loop systems, which recycle water within the industrial facility, and the use of more efficient cooling systems, can reduce the overall demand for fresh water (Zhao et al., 2017). Furthermore, industries are increasingly investing in water-efficient technologies and adopting best practices to reduce water waste and improve sustainability. Collaboration between governments, businesses, and communities is essential for creating policies that promote responsible water use and protect local water resources.

The Role of Technology in Water Conservation

Technological advancements have played a pivotal role in improving water conservation efforts across various sectors. Smart technologies, such as Internet of Things (IoT) sensors, have revolutionized water management by providing real-time data on water usage and enabling better decision-making (Murray, 2016). For example, IoT-based systems in agriculture can monitor soil moisture levels, weather patterns, and crop needs, ensuring that irrigation is optimized for maximum efficiency. Additionally, technologies such as reverse osmosis and desalination have provided alternative solutions to augment freshwater supplies in water-scarce regions (Elimelech & Phillip, 2011). These technologies allow for the conversion of seawater into potable water, although energy requirements and environmental impacts must be carefully managed. The development of low-energy desalination techniques and the integration of renewable energy sources into water management processes are vital for ensuring that these technologies remain sustainable in the long term.

Water Conservation in Urban Areas

Urban areas face unique challenges in managing water resources due to high population densities and increased demand for water. Cities around the world are implementing various strategies to conserve water and reduce urban water footprints. One key approach is the promotion of water-efficient infrastructure, such as rainwater harvesting systems, greywater reuse, and permeable pavements, which allow for water infiltration into the ground (Stefanakis, 2014). Additionally, urban planning initiatives that incorporate green spaces, such as parks and green roofs, can help manage stormwater and improve water quality. Education campaigns in cities also encourage residents to adopt water-saving practices, such as using water-efficient appliances, fixing leaks, and reducing water usage during peak demand periods. As urban populations continue to grow, it is critical to develop integrated water management systems that incorporate both technological innovations and sustainable design principles to ensure the long-term availability of water resources.

Water Conservation and Ecosystem Protection

Water conservation is intrinsically linked to the protection of ecosystems that rely on freshwater resources. Wetlands, rivers, and lakes play crucial roles in supporting biodiversity, regulating water quality, and providing water for agricultural and industrial use. Over-extraction of water for human use can lead to the degradation of these ecosystems, with devastating consequences for wildlife and plant species. For example, the diversion of rivers for irrigation or hydropower projects can disrupt natural habitats and lead to the loss of biodiversity (Postel & Richter, 2012). Protecting ecosystems through water conservation efforts not only ensures that we have access to clean water but also maintains the resilience of natural systems that regulate climate, filter pollutants, and support the livelihoods of millions of people. Effective water management should prioritize ecosystem health, incorporating conservation strategies that safeguard freshwater habitats and the species that depend on them.

Water Conservation and Climate Change

The relationship between water conservation and climate change is complex, as climate change has a direct impact on water availability and quality. Rising temperatures, altered precipitation patterns, and more frequent droughts are expected to exacerbate water scarcity in many regions. Water conservation efforts can help mitigate the effects of climate change by reducing water demand and improving the resilience of water systems. For instance, implementing water-efficient agricultural practices and investing in water-saving technologies can help cope with reduced water availability due to changing weather patterns (Wang et al., 2014). Furthermore, conserving water helps reduce energy consumption associated with water treatment and distribution, which in turn reduces greenhouse gas emissions. Water conservation is thus an important tool in both adapting to and mitigating the impacts of climate change, contributing to overall environmental sustainability.

Corporate Responsibility in Water Conservation

Corporations have a significant impact on global water consumption, particularly in industries such as manufacturing, agriculture, and energy. As part of their corporate social responsibility (CSR) initiatives, many companies are adopting water conservation practices to reduce their environmental footprint and improve sustainability (Schaltegger & Wagner, 2017). Corporations are increasingly focusing on water stewardship, which involves managing water resources in a way that balances business needs with the needs of local communities and ecosystems. Companies are investing in water-efficient technologies, optimizing production processes to reduce water usage, and improving wastewater treatment. Transparent reporting on water usage and conservation efforts is also becoming a standard practice for companies aiming to improve their environmental performance and address the growing concerns over water scarcity.

The Economic Benefits of Water Conservation

Investing in water conservation not only helps preserve a vital resource but also offers significant economic benefits. Efficient water management can lead to cost savings for households, businesses, and governments by reducing water treatment and distribution costs, as well as the costs associated with water shortages (UNEP, 2017). In agriculture, water conservation techniques such as drip irrigation can increase crop yields while reducing water use, thus improving profitability for farmers. In urban areas, water-saving measures can help reduce the strain on municipal water systems, preventing costly infrastructure expansions. Moreover, the economic benefits of water conservation extend to environmental services, as protecting ecosystems that provide clean water can reduce the need for expensive water treatment processes. Overall, water conservation presents a compelling case for both environmental and economic sustainability.

Community-Based Water Conservation Initiatives

Community-based water conservation initiatives play a crucial role in addressing local water challenges and promoting sustainable water use. These initiatives are often driven by local knowledge and involve community members in decision-making processes related to water management. For example, in rural areas, farmers may collaborate to implement water-sharing agreements or collective irrigation systems that improve water use efficiency (Sharma et al., 2015). In urban areas, community-led programs can raise awareness about water conservation, encourage water-efficient behaviors, and advocate for policy changes that promote sustainable water management. By engaging communities in water conservation efforts, these initiatives help build a sense of ownership and responsibility, leading to more effective and lasting results.

Government Policies for Water Conservation

Government policies play a crucial role in fostering water conservation at national and local levels. Governments can implement regulatory frameworks that encourage the sustainable use of water resources. This can include setting water usage standards for industries, promoting water-efficient technologies, and investing in infrastructure that supports water conservation. For example, many countries have implemented pricing strategies that reflect the true cost of water, incentivizing consumers and industries to use water more efficiently (Grafton et al., 2011). Additionally, governments can support research and innovation in water management technologies, such as water recycling and desalination. Public awareness campaigns and education programs can also help foster a culture of water conservation. By creating policies that align with sustainable water use, governments can drive collective action to protect this vital resource.

Water Conservation and Public Awareness

Public awareness is essential for promoting water conservation on a large scale. Educating individuals, communities, and businesses about the importance of water conservation can lead to significant reductions in water consumption. Public campaigns can highlight the value of water and encourage practices such as fixing leaks, using water-efficient appliances, and reducing water usage during peak hours. Schools and educational institutions play a key role in instilling water-saving habits in the younger generation, ensuring that future generations understand the importance of preserving water resources. Social media and digital platforms are increasingly being used to spread awareness and engage people in water conservation activities. Collective action and behavior change, driven by public awareness, are key to achieving sustainable water use.

Corporate Water Stewardship Programs

Corporate water stewardship is gaining importance as more companies recognize the need to manage their water resources responsibly. Companies are adopting strategies to ensure they use water efficiently, reduce waste, and minimize their environmental impact. These programs can include measures such as reducing water consumption in manufacturing processes, recycling wastewater, and engaging in water restoration projects (Van Leeuwen et al., 2017). Leading companies also collaborate with local communities and stakeholders to address water scarcity issues in regions where they operate. Transparent reporting on water usage and the effectiveness of conservation efforts is a critical aspect of corporate water stewardship. By implementing water stewardship programs, companies contribute to long-term water sustainability and demonstrate their commitment to environmental responsibility.

Water Recycling and Reuse

Water recycling and reuse are among the most effective strategies for addressing water scarcity, particularly in urban areas and industries. Water recycling involves treating wastewater to make it suitable for reuse in various applications, such as irrigation, industrial processes, and even potable water. Technologies like membrane filtration, ultraviolet (UV) disinfection, and reverse osmosis are commonly used in water treatment plants to purify wastewater (Abdelrasoul et al., 2017). Reusing water reduces the demand for fresh water and helps mitigate the environmental impact of wastewater disposal. In many regions, especially those facing chronic water shortages, water recycling is a key component of sustainable water management. Encouraging the adoption of water reuse systems in households and industries can significantly reduce water consumption and enhance water security.

Water Conservation in the Hospitality Industry

The hospitality industry, including hotels, resorts, and restaurants, is a significant consumer of water, particularly in areas with high tourist activity. The industry's water usage often exceeds that of residential and commercial sectors, making water conservation a priority for sustainable operations. Many hotels are adopting water-saving measures such as low-flow showerheads, towel reuse programs, and water-efficient landscaping to reduce their overall water consumption (Font & Harris, 2016). Additionally, implementing water recycling and greywater reuse systems can help reduce demand on municipal water supplies. Encouraging guests to adopt water-saving behaviors during their stay, such as limiting water use in the bathroom or reusing towels, can also contribute to reducing water consumption. By embracing sustainable water practices, the hospitality industry can reduce its environmental footprint while helping to preserve water resources.

Global Cooperation in Water Conservation

Water conservation is a global issue that requires cooperation among countries, organizations, and communities to address. Many countries face shared water resources, and effective management of these resources requires transboundary cooperation. International agreements and treaties can help prevent conflicts over water and promote equitable access to freshwater (Zeitoun et al., 2014). For example, the United Nations Watercourses Convention emphasizes the need for cooperation in managing international watercourses to ensure the sustainable use of water resources. Collaborative efforts in water conservation also involve knowledge sharing, capacity building, and joint research initiatives aimed at improving water management practices. Global partnerships, such as the UN's Sustainable Development Goal 6, which focuses on clean water and sanitation for all, provide a framework for international cooperation in water conservation.

Conclusion: The Future of Water Conservation

Water conservation is an essential practice for ensuring the sustainability of global water resources in the face of increasing demand and climate change. Through individual actions, technological innovations, government policies, and corporate stewardship, significant progress can be made in preserving water for future generations. The key to success lies in adopting a holistic approach that includes water-saving practices in all sectors, from agriculture to industry, and fostering public awareness about the importance of water conservation. As the global population grows and climate challenges intensify, water conservation must remain a top priority for governments, businesses, and individuals alike. By working together, we can secure the future of water resources and promote a sustainable, water-efficient world.

Chapter 8: Waste Management: Reducing and recycling

Introduction to Waste Management

Waste management is a critical aspect of environmental conservation, aiming to minimize the environmental impacts of waste and promote sustainable practices. The improper handling of waste can lead to a range of issues, including pollution, resource depletion, and health hazards. Therefore, waste management focuses on reducing waste generation, reusing materials, and recycling as much as possible. Reducing waste at the source is the most effective way to manage waste, as it limits the need for disposal and the environmental consequences associated with it (Hester & Harrison, 2018). Recycling, on the other hand, involves the reprocessing of materials into new products, reducing the demand for raw materials and saving energy. Proper waste management not only helps mitigate environmental pollution but also offers economic benefits, such as job creation in the recycling industry. As global populations grow and consumption increases, efficient waste management strategies are becoming more crucial than ever.

The Waste Hierarchy: Reducing, Reusing, and Recycling

The waste hierarchy is a guiding framework for managing waste, with a focus on reducing waste generation and minimizing environmental harm. It consists of three key strategies: reduce, reuse, and recycle. The first priority in the hierarchy is to reduce waste at the source. This involves making changes in product design, packaging, and consumption patterns to minimize waste generation (Kinnunen et al., 2014). Reuse is the second priority, which involves finding ways to extend the life of products by repurposing them or using them for alternative functions. The third priority, recycling, involves processing waste materials into new products, conserving raw materials, and reducing energy use. Although recycling is an essential part of waste management, it is most effective when combined with reducing and reusing, as it helps create a circular economy, where materials are kept in use for as long as possible (Geissdoerfer et al., 2017).

Waste Reduction Strategies

Waste reduction involves minimizing the amount of waste generated by individuals, businesses, and industries. One effective strategy is product redesign, where manufacturers create products with less packaging, fewer toxic materials, and better durability (Kumar & Saini, 2015). Another strategy is the adoption of sustainable practices by consumers, such as reducing food waste by purchasing only what is needed and using items for multiple purposes. Industrial waste reduction can also be achieved by optimizing production processes, reducing material waste, and implementing energy-efficient technologies. Moreover, waste reduction can be promoted through policy measures, such as implementing extended producer responsibility (EPR) laws that require producers to take responsibility for the entire lifecycle of their products, including their disposal (Tait, 2019). By embracing waste reduction strategies, both individuals and industries can contribute to a more sustainable future by reducing their environmental footprint.

Recycling: The Importance and Process

Recycling is the process of converting waste materials into new products, which helps reduce the consumption of fresh raw materials, lower energy usage, and decrease pollution. The recycling process involves several stages, including collection, sorting, cleaning, and processing. Materials such as paper, glass, metal, and plastics are commonly recycled into new products. For example, aluminum cans can be melted and refashioned into new cans or other aluminum products, while paper can be recycled into new paper products or insulation materials (Bajada et al., 2017). The benefits of recycling extend beyond waste reduction. By reducing the demand for raw materials, recycling helps conserve natural resources and reduces the environmental impact of resource extraction. It also saves energy, as producing recycled products typically requires less energy than producing new products from virgin materials (Dijkgraaf & Vollebergh, 2004). To maximize recycling efforts, individuals and businesses must follow proper waste segregation practices to ensure materials are clean and suitable for recycling.

The Role of Composting in Waste Management

Composting is the natural process of recycling organic waste, such as food scraps, yard trimmings, and agricultural waste, into nutrient-rich soil. Composting reduces the amount of organic waste sent to landfills, where it would otherwise decompose anaerobically, producing methane – a potent greenhouse gas (Koch & Schwaiger, 2015). By turning organic waste into compost, valuable soil amendments can be produced that improve soil health, enhance water retention, and reduce the need for chemical fertilizers. Community-based composting programs are increasingly being implemented in urban areas, where food waste is a significant contributor to landfill mass (Sanchez et al., 2017). These programs encourage households and businesses to separate their organic waste and participate in the composting process. Composting is an integral part of waste management systems, as it contributes to reducing landfill waste, enhancing soil fertility, and reducing environmental impacts.

Electronic Waste Management

The rapid growth of electronic devices has led to an increase in electronic waste (e-waste), which presents significant challenges for waste management systems. E-waste contains valuable materials, such as gold, copper, and silver, which can be recovered through recycling processes. However, improper disposal of e-waste, particularly in landfills or informal recycling sectors, can lead to environmental pollution due to the release of toxic substances like lead, mercury, and cadmium (Ongondo et al., 2011). The management of e-waste involves the safe collection, recycling, and disposal of electronics. Many countries have implemented regulations that require manufacturers to take responsibility for the recycling of their products through programs such as take-back schemes and producer responsibility programs (Forti et al., 2020). As technology continues to evolve, effective e-waste management will remain a critical aspect of global waste management strategies.

Waste-to-Energy (WTE) Technologies

Waste-to-energy (WTE) technologies are an emerging approach to managing waste while simultaneously generating energy. These technologies involve converting non-recyclable waste into usable energy, typically through processes such as combustion, gasification, or anaerobic digestion. WTE plants help reduce the volume of waste sent to landfills while also providing a renewable energy source (Minghua et al., 2016). The energy produced can be used for electricity generation or heating purposes, providing an alternative to fossil fuels. While WTE has the potential to contribute to energy security and waste reduction, it also has environmental impacts, including air pollution and carbon emissions. Therefore, careful regulation and integration with other waste management strategies, such as recycling and composting, are essential to ensure that WTE technologies contribute to a sustainable waste management system (Nair et al., 2017). The growing interest in WTE technologies reflects a broader shift towards utilizing waste as a resource rather than simply a problem to be disposed of.

The Importance of Reducing Single-Use Plastics

Single-use plastics are one of the most significant contributors to global waste, as they are designed to be used once and discarded. These plastics, which include items such as straws, cutlery, and plastic bags, often end up in landfills, where they take hundreds of years to decompose, or worse, in the oceans, where they harm marine life (Jambeck et al., 2015). The widespread use of single-use plastics has prompted calls for stricter regulations and efforts to reduce plastic consumption. One of the key strategies for addressing this issue is encouraging the use of reusable alternatives, such as cloth bags, metal straws, and bamboo cutlery. Additionally, extended producer responsibility (EPR) policies that hold manufacturers accountable for the entire lifecycle of their products, including waste disposal, can incentivize companies to reduce plastic packaging and design products with recyclability in mind (Schroeder et al., 2019). Reducing single-use plastics is critical for reducing overall waste generation and mitigating the environmental impact of plastic pollution.

The Role of Extended Producer Responsibility (EPR)

Extended Producer Responsibility (EPR) is an environmental policy approach that holds manufacturers accountable for the entire lifecycle of their products, including post-consumer waste. Under EPR, producers are responsible for the collection, recycling, and disposal of their products once they reach the end of their useful life (Bjørn et al., 2018). This policy incentivizes companies to design products that are easier to recycle, use fewer toxic materials, and have a smaller environmental footprint. EPR has been successfully implemented in various countries, particularly in the management of electronic waste (e-waste) and packaging materials. For example, the European Union's Waste Electrical and Electronic Equipment (WEEE) Directive mandates that manufacturers take responsibility for the recycling of e-waste. Similarly, some countries have enacted laws requiring manufacturers to take back plastic packaging and invest in recycling infrastructure. By shifting the responsibility to producers, EPR encourages the development of a circular economy and reduces the environmental impact of waste.

Circular Economy and Waste Management

The concept of a circular economy is central to modern waste management strategies, emphasizing the importance of reusing and recycling materials to create a closed-loop system. In a circular economy, products are designed for durability, repairability, and recyclability, and waste is minimized by extending the lifespan of materials (Geissdoerfer et al., 2017). This contrasts with the traditional linear economy, which follows a "take-make-dispose" model, where resources are extracted, products are manufactured, used, and then discarded as waste. A circular economy promotes the idea that materials should circulate within the system for as long as possible, reducing the need for new resources and minimizing waste sent to landfills. Implementing circular economy principles in waste management can lead to substantial environmental, economic, and social benefits, such as resource conservation, job creation, and reduced pollution (Kirchherr et al., 2018). To achieve a circular economy, stakeholders must collaborate across sectors, including manufacturing, government, and consumers, to design sustainable products and improve waste management practices.

Waste Management in Developing Countries

Waste management in developing countries presents unique challenges, including limited infrastructure, inadequate waste collection services, and insufficient waste segregation practices. In many cases, waste is either openly dumped or burned, leading to environmental contamination and health risks. However, some developing countries are making strides in implementing sustainable waste management practices. For example, some countries in Africa have adopted informal waste recycling systems, where local communities collect and recycle materials like plastics and metals for resale. While these efforts can help alleviate waste problems, they often lack proper safety and environmental regulations (Wilson et al., 2015). The international community, as well as national governments, must provide technical assistance, financial support, and training to improve waste management practices in developing countries. Strengthening waste management infrastructure, promoting waste segregation, and encouraging recycling are crucial steps to reducing environmental pollution and improving public health in these regions.

The Impact of Landfills on the Environment

Landfills have long been the default solution for waste disposal, but they are associated with significant environmental problems. One of the primary concerns is the release of methane, a potent greenhouse gas that is produced when organic waste decomposes anaerobically in landfills (Czepiel et al., 2003). In addition to methane, landfills can contaminate soil and groundwater with toxic substances from leachate, the liquid that forms when waste breaks down. This can result in the pollution of nearby water sources, posing a threat to both ecosystems and human health. Furthermore, landfills take up valuable land space and contribute to urban sprawl. To mitigate the environmental impact of landfills, modern waste management systems incorporate strategies such as landfill gas capture and leachate treatment, as well as promoting recycling and waste diversion to reduce the amount of waste sent to landfills (Zhang et al., 2017).

Greenhouse Gas Emissions and Waste Management

Waste management has a significant impact on greenhouse gas (GHG) emissions, which are a major driver of climate change. Waste management processes, such as landfilling and waste incineration, release gases like carbon dioxide (CO2), methane (CH4), and nitrous oxide (N2O) into the atmosphere (Fiedler et al., 2017).

Methane, produced during the decomposition of organic waste in landfills, is particularly concerning because it has a global warming potential that is 25 times greater than that of CO2 over a 100-year period. To reduce GHG emissions from waste, waste management systems need to prioritize the reduction of landfill disposal and the promotion of recycling and composting. For example, by diverting organic waste from landfills through composting, methane emissions can be significantly reduced (Møller et al., 2019). Similarly, recycling reduces the need for raw material extraction and minimizes the energy consumption associated with producing new products from virgin materials. Transitioning to more sustainable waste management practices is essential for mitigating climate change and reducing GHG emissions.

The Benefits of Sustainable Waste Management

Sustainable waste management has far-reaching benefits, both for the environment and for society. By reducing, reusing, and recycling materials, waste management systems help conserve natural resources, reduce pollution, and mitigate climate change. For example, recycling conserves raw materials and reduces energy consumption, as producing recycled products typically requires less energy than manufacturing products from virgin materials (Dijkgraaf & Vollebergh, 2004). Waste diversion from landfills can also reduce greenhouse gas emissions, as landfill-bound waste contributes to methane production (Fiedler et al., 2017). Furthermore, sustainable waste management practices can generate economic benefits by creating jobs in the recycling, composting, and waste-to-energy sectors (Bajada et al., 2017). In addition, reducing waste at the source encourages innovation in product design, which leads to more durable and recyclable products. Ultimately, transitioning to sustainable waste management systems is essential for creating a circular economy, where waste is minimized, and resources are used more efficiently.

Waste Management and Public Health

Waste management is closely linked to public health, as improper waste disposal can lead to various health risks. For instance, open burning of waste releases toxic substances, including dioxins and particulate matter, which can cause respiratory problems and other health issues (Chen et al., 2017). Similarly, poorly managed landfills can attract disease-carrying pests, such as rodents and mosquitoes, which can spread diseases like malaria and cholera. Waste management systems that prioritize proper waste segregation, recycling, and safe disposal can significantly reduce these health risks. Additionally, waste-to-energy technologies can mitigate the harmful effects of waste incineration by capturing and treating harmful emissions (Nair et al., 2017). Finally, waste management systems that focus on composting organic waste help reduce exposure to toxic substances and prevent contamination of water and soil. Proper waste management practices, therefore, not only protect the environment but also play a crucial role in safeguarding public health.

Education and Awareness in Waste Management

Education and public awareness are essential components of effective waste management strategies. By raising awareness about the importance of waste reduction, recycling, and composting, individuals can make informed decisions about their consumption patterns and waste disposal behaviors (Vargas et al., 2018). Schools, community groups, and governments play a key role in educating the public about the benefits of sustainable waste management and the negative environmental impacts of improper waste disposal. For example, educational campaigns can promote the idea of reducing single-use plastics, encourage composting at home, and provide information about local recycling programs. Additionally, businesses can incorporate sustainability practices into their operations, such as offering incentives for customers to recycle or reducing the amount of packaging used for products. Ultimately, fostering a culture of waste-consciousness and encouraging sustainable behaviors will help reduce the environmental impact of waste and ensure that waste management systems are more efficient.

The Role of Technology in Waste Management

Technology plays a crucial role in improving waste management systems and increasing their efficiency. Advancements in waste processing technologies, such as automated sorting and waste-to-energy plants, have helped streamline waste handling, reduce contamination, and increase recycling rates (Ongondo et al., 2011). For example, smart waste management systems utilize sensors and real-time data to monitor waste levels in bins and optimize collection routes, reducing fuel consumption and operational costs (Raut et al., 2019). Additionally, new technologies in recycling, such as chemical recycling, are being explored to process difficult-to-recycle materials like mixed plastics and textiles. The integration of artificial intelligence (AI) in waste sorting can also improve the efficiency and accuracy of sorting recyclable materials (Lin et al., 2020). As technology continues to evolve, it holds the potential to revolutionize waste management practices, making them more sustainable, cost-effective, and environmentally friendly.

Waste Management and the Circular Economy

The circular economy concept emphasizes reducing waste and promoting the reuse, repair, and recycling of materials to create a closed-loop system. In a circular economy, waste is seen as a resource rather than a problem, with the goal of maintaining the value of products and materials in the economy for as long as possible (Geissdoerfer et al., 2017). By embracing circular economy principles, businesses can reduce the environmental impacts of production and consumption, conserve resources, and reduce waste generation. For example, companies can design products for durability, use recyclable materials, and encourage customers to return used products for recycling or refurbishment. Governments can support the transition to a circular economy through policy measures, such as waste reduction targets, EPR schemes, and incentives for businesses that adopt sustainable practices (Kirchherr et al., 2018). The shift to a circular economy requires collaboration across sectors, including manufacturing, recycling, and waste management, to create a sustainable and resilient system.

Global Efforts in Waste Management

Globally, waste management practices vary, with developed countries often having more established systems compared to developing nations. However, waste management is a growing concern worldwide, particularly in urban areas where population growth and industrialization lead to higher waste generation (Saha & Banerjee, 2020). The adoption of sustainable waste management practices is crucial to addressing the environmental challenges posed by urbanization and industrial waste. International organizations, such as the United Nations (UN), are actively working to promote waste management best practices, with initiatives like the Sustainable Development Goals (SDGs) aiming to reduce waste generation and encourage recycling by 2030 (UNEP, 2018). Countries like Sweden, Germany, and South Korea are leading the way in waste management through strict regulations, extensive recycling programs, and the promotion of waste-to-energy technologies (Schweizer, 2017). Meanwhile, developing nations are facing significant challenges in waste management, including a lack of infrastructure, limited resources, and growing waste volumes. By fostering international collaboration, sharing knowledge, and providing technical support, it is possible to help less developed countries transition to more sustainable waste management systems.

The Future of Waste Management

The future of waste management will be shaped by advances in technology, the growing need for sustainability, and changes in consumer behavior. A key aspect of future waste management systems will be the increased adoption of circular economy practices, where materials are reused and recycled rather than disposed of in landfills (Ellen MacArthur Foundation, 2019). Emerging technologies, such as artificial intelligence (AI) for waste sorting and blockchain for waste tracking, are expected to play a significant role in increasing the efficiency and transparency of waste management systems (Cox et al., 2018). Additionally, waste-to-energy technologies, including anaerobic digestion and waste pyrolysis, offer promising solutions for converting waste into useful energy while reducing landfill use (Kuo et al., 2019). Governments will need to continue implementing policies that promote recycling, waste reduction, and the sustainable management of resources. Public education campaigns will also play an important role in encouraging more sustainable consumption patterns and waste disposal behaviors. Ultimately, the transition toward a more sustainable waste management system will require coordinated efforts from governments, businesses, and individuals to reduce waste, minimize environmental impact, and preserve valuable resources for future generations.

Chapter 9: Ocean Health: Protecting marine life

Introduction to Ocean Health

The health of the world's oceans is a critical aspect of environmental sustainability. Oceans cover over 70% of the Earth's surface and play a central role in regulating the climate, providing food, and supporting biodiversity (UNESCO, 2020). Marine life, ranging from microscopic plankton to massive whales, is an integral part of ocean ecosystems. However, human activities have significantly impacted ocean health, leading to threats such as pollution, overfishing, and climate change (Hoegh-Guldberg et al., 2019). Protecting marine life is essential for maintaining these ecosystems' functions, which include carbon sequestration, food security, and coastal protection. As human populations grow, it becomes even more crucial to address the factors harming oceans and implement strategies to protect marine life. This paper explores the threats to ocean health, the importance of marine biodiversity, and potential solutions to preserve ocean ecosystems.

The Importance of Marine Life

Marine life forms the foundation of ocean ecosystems, contributing to biodiversity and sustaining ecological balance. The oceans host an estimated 230,000 species of marine organisms, ranging from microscopic plankton to large mammals like whales (Moreno et al., 2017). These species play vital roles, such as providing food for other animals, cycling nutrients, and supporting the production of oxygen through photosynthesis. Marine life is not only crucial for the environment but also for human societies. Fisheries provide a significant portion of protein for billions of people, while coral reefs, often referred to as the "rainforests of the sea," support a vast array of species and protect coastal communities from storm surges (Moberg & Folke, 1999). Thus, the health of marine life is directly linked to the well-being of both the environment and human populations. Protecting marine life means ensuring the continued functioning of these essential services.

Major Threats to Ocean Health

Several human-driven activities are causing significant harm to ocean ecosystems. Pollution is one of the primary threats to marine life, with plastics being the most visible and harmful pollutant. Each year, millions of tons of plastic waste end up in the oceans, affecting marine animals that ingest or become entangled in plastic debris (Jambeck et al., 2015). Overfishing is another major concern, leading to the depletion of fish stocks and the disruption of marine food webs (Pauly et al., 2002). In addition, climate change exacerbates ocean health problems by increasing sea temperatures, acidifying ocean waters, and causing coral bleaching events. These factors combined create a perfect storm of threats, making it urgent to address the root causes of ocean degradation and implement effective conservation strategies.

The Role of Marine Ecosystems in Climate Regulation

Marine ecosystems play an indispensable role in regulating the global climate. Oceans act as a major carbon sink, absorbing a significant portion of the carbon dioxide (CO2) emitted into the atmosphere (Sabine et al., 2004). Phytoplankton, the tiny marine plants that live in the upper layers of the ocean, absorb CO2 during photosynthesis and release oxygen, thus helping mitigate the effects of climate change. Coral reefs, seagrasses, and mangroves are also important in carbon sequestration, with coastal ecosystems storing vast amounts of carbon in their soils (Fourqurean et al., 2012). The degradation of these ecosystems, however, reduces their capacity to sequester carbon and exacerbates climate change. Protecting these ecosystems is therefore vital for maintaining their role in climate regulation and ensuring that oceans continue to help mitigate global warming.

Ocean Acidification and Its Impact on Marine Life

Ocean acidification is one of the most significant threats to marine ecosystems resulting from human activities. As atmospheric CO2 levels rise, more CO2 is absorbed by the oceans, which reacts with seawater to form carbonic acid, lowering the pH of ocean waters (Doney et al., 2009). This acidification process disrupts the ability of marine organisms, such as corals, mollusks, and certain plankton species, to build calcium carbonate shells and skeletons (Orr et al., 2005). The loss of these organisms has cascading effects on marine food webs, as many species rely on them as a source of food or habitat. Ocean acidification also threatens the health of coral reefs, which are vital for biodiversity and coastal protection. Efforts to combat ocean acidification include reducing carbon emissions and protecting marine ecosystems to enhance their resilience.

The Impact of Overfishing on Marine Ecosystems

Overfishing is one of the most critical threats to ocean health. It depletes fish stocks at unsustainable rates, disrupting marine food webs and threatening biodiversity. According to the Food and Agriculture Organization (FAO), about 34% of global fish stocks are overexploited, and many others are nearing overfished status (FAO, 2020). The removal of large numbers of fish disrupts predator-prey relationships and reduces the availability of food for other marine animals. This, in turn, affects not only the health of marine life but also the livelihoods of millions of people who rely on fishing for food and income. Furthermore, overfishing leads to the destruction of important habitats, such as coral reefs and seagrass meadows, as fishing boats and gear damage the ocean floor (Kerr, 2016). Implementing sustainable fishing practices, such as setting catch limits and using eco-friendly fishing gear, is essential to protect marine biodiversity and ensure the long-term health of the oceans.

Plastic Pollution and Its Threat to Marine Life

Plastic pollution is one of the most visible and harmful threats to ocean health. It is estimated that around 8 million tons of plastic enter the oceans every year, causing widespread damage to marine life (Jambeck et al., 2015). Marine animals, including fish, seabirds, and marine mammals, often mistake plastic debris for food, leading to ingestion, choking, and death. Additionally, plastics break down into microplastics, which are ingested by even the smallest marine organisms, entering the food chain and potentially affecting human health (Rochman et al., 2013). Plastic waste also damages marine habitats, particularly coral reefs, where plastic debris smothers the delicate organisms that form these ecosystems (Keshavarz et al., 2019). Reducing plastic pollution requires concerted efforts from governments, industries, and individuals, including reducing plastic production, improving waste management, and promoting recycling.

Coral Reefs: A Biodiversity Hotspot

Coral reefs are among the most biologically diverse ecosystems on Earth, providing habitat for a wide variety of marine species. These ecosystems support over 25% of all marine life, despite covering less than 0.1% of the ocean's surface (Wilkinson, 2008). However, coral reefs are increasingly threatened by human activities. Climate change is one of the most significant threats, causing rising sea temperatures that lead to coral bleaching, a process in which corals expel the symbiotic algae that provide them with energy and color (Hughes et al., 2017). Coral reefs are also impacted by ocean acidification, overfishing, and pollution. The degradation of coral reefs has serious consequences, as they provide critical services such as coastal protection, carbon sequestration, and sustenance for coastal communities. Protecting coral reefs requires efforts to reduce greenhouse gas emissions, establish marine protected areas, and regulate harmful fishing practices.

Marine Protected Areas (MPAs) as a Conservation Tool

Marine protected areas (MPAs) are designated regions in the ocean where human activities are regulated to protect marine ecosystems. MPAs can vary in size and the degree of protection they offer, ranging from no-take zones, where fishing and other extractive activities are prohibited, to areas with more flexible management strategies (Costello et al., 2010). MPAs play a crucial role in conserving marine biodiversity by providing safe havens for marine life to thrive, rebuild populations, and recover from the pressures of overfishing and habitat destruction (Beger et al., 2015). Research has shown that MPAs can help replenish fish stocks, restore coral reefs, and improve the overall health of marine ecosystems (Edgar et al., 2014). Expanding and effectively managing MPAs is essential for the long-term protection of marine life and the overall health of the oceans.

The Role of Sustainable Fisheries Management

Sustainable fisheries management is a key component of protecting marine life and ensuring the long-term health of ocean ecosystems. Overfishing, bycatch, and destructive fishing practices are major threats to marine biodiversity (Pauly et al., 2002). Implementing sustainable fishing practices involves setting catch limits, regulating fishing gear, and reducing bycatch to prevent the depletion of fish stocks and the destruction of marine habitats. One promising approach is the use of catch share programs, which allocate a specific portion of the total allowable catch to individual fishers or communities, incentivizing them to manage fish stocks responsibly (Costello et al., 2008). In addition, promoting the use of sustainable seafood, as indicated by certification programs such as the Marine Stewardship Council (MSC), can help consumers make informed choices and support responsible fisheries.

The Impact of Climate Change on Marine Life

Climate change is one of the most significant challenges facing ocean health today. Rising global temperatures are leading to increased sea surface temperatures, which in turn are causing coral bleaching and the disruption of marine ecosystems. As oceans warm, species that are adapted to specific temperature ranges are being forced to migrate or face extinction (Pörtner et al., 2014). Additionally, climate change is contributing to ocean acidification, which threatens the survival of calcifying organisms, such as corals and shellfish, by hindering their ability to form shells and skeletons (Orr et al., 2005). The combined effects of warming waters, acidification, and habitat loss are creating an existential threat to many marine species. Addressing climate change by reducing greenhouse gas emissions and implementing adaptation strategies is vital to protect marine ecosystems and biodiversity.

Addressing Marine Pollution Through Policy and Innovation

Efforts to reduce marine pollution require coordinated global action and innovative solutions. International policies such as the United Nations' Sustainable Development Goal 14 (SDG 14), which aims to conserve and sustainably use the oceans, seas, and marine resources, have gained momentum in recent years (UN, 2015). Policies focused on reducing plastic waste, implementing marine protected areas, and enforcing stricter regulations on industrial pollution are key components of this global effort. Technological innovation also plays a crucial role in mitigating marine pollution. For example, the development of biodegradable plastics and advancements in waste management technologies can significantly reduce the amount of plastic waste entering the oceans (Jambeck et al., 2015). Collaborative initiatives, such as the Global Partnership on Marine Litter, are also working to tackle marine pollution through the exchange of knowledge, resources, and best practices (UNEP, 2018). These combined efforts are essential for protecting ocean health and preserving marine biodiversity.

The Importance of Public Awareness in Ocean Conservation

Raising public awareness about the importance of ocean conservation is essential for the protection of marine life. Educational programs and campaigns that highlight the impacts of human activities, such as plastic pollution, overfishing, and habitat destruction, can help foster a sense of responsibility and stewardship among individuals and communities. Programs like "Ocean Literacy," which teach the general public about the significance of oceans and marine life, are critical in promoting conservation efforts (National Oceanic and Atmospheric Administration, 2013). In addition, social media platforms have become powerful tools in spreading awareness about ocean health issues and mobilizing support for marine conservation initiatives. Public pressure can lead to the implementation of policies, the reduction of plastic consumption, and the adoption of sustainable practices, all of which contribute to the preservation of ocean ecosystems.

The Role of Community-Based Conservation

Community-based conservation (CBC) initiatives are increasingly recognized as effective strategies for protecting marine ecosystems and empowering local populations. CBC projects engage local communities in the management of marine resources, combining traditional knowledge with scientific research to create sustainable solutions (Berkes, 2009). By involving local people in conservation efforts, these initiatives help promote stewardship and ensure that conservation measures are culturally appropriate and effective. For example, community-managed marine protected areas in the Pacific Islands have led to the restoration of fish stocks and the protection of coral reefs (Govan et al., 2011). Supporting and expanding CBC efforts worldwide is essential to building resilient marine ecosystems and communities that are better equipped to cope with environmental challenges.

The Role of Ocean Restoration Projects

Ocean restoration projects are gaining traction as vital tools in reversing some of the damage caused by human activities. These projects aim to restore degraded marine habitats, such as coral reefs, mangroves, and seagrasses, which are critical for biodiversity and coastal protection. Restoration techniques include coral planting, mangrove reforestation, and the restoration of seagrass meadows (Schläppy et al., 2013). Successful restoration not only enhances the resilience of marine ecosystems but also provides significant benefits to local communities, such as increased fish stocks and improved coastal protection from storms and erosion (Bayraktarov et al., 2016). Expanding ocean restoration efforts is key to rebuilding marine ecosystems and enhancing their capacity to support biodiversity.

The Need for International Cooperation in Marine Conservation

Marine conservation is a global challenge that requires international cooperation to be effective. The oceans do not recognize national borders, and the health of marine ecosystems is impacted by activities in multiple countries. Global agreements, such as the Convention on Biological Diversity (CBD) and the United Nations Convention on the Law of the Sea (UNCLOS), are instrumental in coordinating conservation efforts across nations (Lubchenco et al., 2003). International cooperation also facilitates the sharing of resources, expertise, and data, which is critical for tackling issues such as overfishing, marine pollution, and the impacts of climate change. Collaborative efforts, including the creation of transboundary marine protected areas and joint research initiatives, are essential for preserving the health of the world's oceans and marine life.

Green Marine Technologies for Sustainable Ocean Management

Innovative marine technologies can significantly contribute to sustainable ocean management. Green marine technologies, such as eco-friendly fishing gear, renewable energy solutions like offshore wind farms, and low-impact shipping technologies, are designed to reduce the environmental footprint of human activities on the oceans (Ehlers et al., 2019). For example, the development of biodegradable fishing nets can help reduce the number of marine debris, while the use of satellite monitoring systems allows for more efficient management of marine protected areas (Bergström et al., 2014). By investing in and scaling up the use of these technologies, the global community can create a more sustainable future for both the oceans and the industries that depend on them.

The Future of Ocean Health

The future of ocean health depends on our ability to reduce human impacts and restore the balance of marine ecosystems. Climate change, pollution, and overfishing remain major threats, but the growing global awareness and commitment to ocean conservation provide hope for the future. As science, policy, and innovation converge, new solutions for protecting marine life are emerging. For example, advances in ocean-based carbon capture technologies and the growing trend of sustainable seafood consumption are offering pathways toward a more sustainable future for both marine life and human societies (Gierus, 2020). Global cooperation and action are essential to ensure the continued vitality of our oceans.

The Role of Education and Advocacy in Marine Protection

Education and advocacy are essential for driving positive change in ocean conservation. Through public outreach programs, advocacy groups and educational institutions play a vital role in raising awareness of ocean health issues. Organizations like Oceana and Greenpeace work to inform the public, advocate for policy changes, and encourage sustainable consumer choices (Oceana, 2020). Educational programs, from school curricula to community outreach, empower individuals to make a difference by reducing their carbon footprints, minimizing plastic use, and supporting conservation initiatives. Advocacy at local, national, and international levels can lead to the establishment of marine protected areas, stricter regulations on fishing practices, and better waste management systems, all contributing to the protection of marine life.

Conclusion: Protecting Our Oceans for Future Generations

In conclusion, the protection of ocean health is critical for ensuring the survival of marine life and the sustainability of ocean ecosystems. The oceans provide numerous essential services, including climate regulation, food security, and biodiversity support. However, human activities are increasingly putting these vital ecosystems at risk. Efforts to reduce pollution, manage fisheries sustainably, and combat climate change are key to protecting marine life. International cooperation, community-based conservation, innovative technologies, and public awareness are essential components of a global effort to safeguard ocean health. By taking action today, we can ensure that future generations will inherit a healthy and vibrant ocean.

Chapter 10: Urban Sustainability: Greening our cities

Introduction to Urban Sustainability

Urban sustainability refers to the creation and maintenance of cities that are environmentally responsible, socially inclusive, and economically viable for future generations (World Bank, 2017). With over half of the global population now living in urban areas, the importance of sustainable cities has never been more pressing (UN, 2018). Urban areas consume significant resources and produce large amounts of waste, yet they also hold the potential for innovative solutions to address environmental challenges. Greening our cities is a fundamental aspect of urban sustainability, involving the incorporation of nature-based solutions into urban planning and development (Tzoulas et al., 2007). This can range from increasing urban green spaces to implementing sustainable transportation systems and energy-efficient buildings. By prioritizing green practices, cities can reduce their ecological footprint while improving the quality of life for residents.

The Need for Sustainable Urban Development

Sustainable urban development is essential for addressing global challenges such as climate change, resource depletion, and biodiversity loss. Cities are responsible for the majority of greenhouse gas emissions, and with rapid urbanization, the demand for energy, water, and land continues to grow (Seto et al., 2014). To achieve urban sustainability, cities must reduce their environmental impact by incorporating eco-friendly practices into every aspect of urban life. This includes enhancing energy efficiency, reducing waste generation, and creating green spaces to improve air quality and biodiversity. Furthermore, sustainable cities should prioritize social inclusion, ensuring that all residents, regardless of income or background, have access to affordable housing, healthcare, and transportation (Satterthwaite, 2016). In this context, urban sustainability is not only about environmental stewardship but also about creating cities that are equitable and resilient to future challenges.

Greening Urban Spaces: Benefits of Green Infrastructure

Green infrastructure plays a critical role in enhancing urban sustainability by integrating nature into the built environment. Green spaces, such as parks, green roofs, and urban forests, not only provide aesthetic and recreational benefits but also help mitigate environmental problems such as air pollution, flooding, and heat islands (Kabisch et al., 2017). For example, trees and vegetation help absorb carbon dioxide and release oxygen, improving air quality and reducing the urban heat island effect (Bowler et al., 2010). Urban green spaces also support biodiversity by providing habitats for wildlife and promoting ecosystem services such as pollination and water filtration. Investing in green infrastructure can help create more resilient cities by reducing the vulnerability of urban areas to extreme weather events and climate change (Pauleit et al., 2019).

Sustainable Energy Solutions for Cities

Energy consumption is a major contributor to urban environmental impacts. Sustainable energy solutions are essential for reducing cities' carbon footprints and ensuring long-term urban sustainability. Transitioning to renewable energy sources such as solar, wind, and geothermal can significantly reduce greenhouse gas emissions and decrease reliance on fossil fuels (Barton et al., 2017). Many cities around the world are investing in renewable energy technologies and infrastructure to transition toward greener energy systems. Additionally, energy-efficient buildings, smart grids, and energy storage technologies are helping cities optimize energy use and reduce waste (Rosato et al., 2019). By adopting sustainable energy practices, cities can improve air quality, reduce energy costs, and create a cleaner, more resilient environment.

Sustainable Transportation Systems

Transportation is another major contributor to environmental degradation in urban areas, with cars and trucks being the leading sources of air pollution and greenhouse gas emissions. To achieve urban sustainability, cities must transition to sustainable transportation systems that prioritize public transit, cycling, walking, and electric vehicles (Banister, 2018). Public transportation systems, such as buses, trains, and subways, can significantly reduce the number of private vehicles on the road, thereby lowering emissions and reducing traffic congestion (Givoni, 2016). Moreover, integrating cycling and walking infrastructure into urban design encourages healthier and more sustainable modes of transport. Electric vehicles, when powered by renewable energy, offer an alternative to traditional fossil fuel-powered vehicles and can reduce urban air pollution (Sovacool et al., 2018). Investing in sustainable transportation systems is crucial for creating livable, green cities.

The Role of Urban Agriculture in Sustainability

Urban agriculture, the practice of cultivating food within cities, plays a vital role in promoting urban sustainability. By growing food locally, cities can reduce the carbon footprint associated with food transportation, decrease reliance on industrial agriculture, and improve food security (Zezza & Tasciotti, 2010). Urban farming also provides residents with access to fresh produce, which is particularly important in areas where food deserts exist (Morgan, 2009). In addition, urban agriculture can contribute to green infrastructure by transforming vacant lots or underutilized spaces into productive agricultural areas (McClintock, 2014). Furthermore, urban farming promotes community engagement, strengthens local economies, and enhances social well-being by fostering collaboration and connections among urban dwellers. These benefits make urban agriculture a key component of sustainable cities.

Waste Management and Recycling in Urban Areas

Waste management is one of the most pressing challenges faced by urban areas today. Cities generate large amounts of waste, and without effective management systems, this waste can overwhelm landfills and pollute the environment. Sustainable waste management strategies, including waste reduction, recycling, and composting, are essential for reducing the environmental impact of urbanization (Wilson et al., 2015). Recycling programs help divert materials such as plastics, metals, and paper from landfills, reducing the need for virgin resources and minimizing pollution (Gupta & Khurana, 2017). Composting organic waste creates nutrient-rich soil that can be used in urban agriculture or for landscaping, further contributing to the sustainability of cities (Ganguly et al., 2018). Encouraging waste reduction at the individual and community levels can also reduce the overall amount of waste generated, making cities more sustainable and environmentally friendly.

Water Management and Conservation in Cities

Water scarcity is a growing concern in many urban areas, particularly in regions facing droughts and climate change impacts. Sustainable water management is crucial for ensuring that urban populations have access to clean, safe, and affordable water. Efficient water use, along with effective wastewater treatment and recycling, is essential for reducing water consumption and minimizing the environmental impact of urbanization (Gleick, 2014). One key strategy is the implementation of water-efficient technologies in homes, businesses, and public infrastructure, such as low-flow faucets and smart irrigation systems (Rossi et al., 2016). Moreover, cities can promote rainwater harvesting to supplement water supply and reduce dependence on external water sources (Zhao et al., 2015). Urban planners also need to prioritize the protection of natural water bodies and wetlands, which provide essential ecosystem services such as water filtration and flood control (McDonald et al., 2014). Integrating sustainable water practices is essential for the long-term resilience of urban areas.

Building Resilient and Sustainable Infrastructure

Resilient infrastructure is a key component of urban sustainability. As cities face growing threats from climate change, such as extreme weather events, rising sea levels, and flooding, the need for robust infrastructure that can withstand these challenges becomes ever more urgent (Hallegatte et al., 2013). Sustainable building materials, such as recycled steel, bamboo, and green concrete, can help reduce the environmental impact of construction while ensuring the durability of structures (Kibert, 2016). Additionally, green roofs and walls not only reduce energy consumption but also contribute to urban biodiversity (Oberndorfer et al., 2007). The integration of renewable energy sources and energy-efficient technologies into buildings helps reduce carbon emissions and energy consumption. Furthermore, cities can incorporate climate adaptation strategies, such as flood prevention systems and stormwater management solutions, to enhance their resilience and sustainability (O'Brien & Leichenko, 2009).

The Role of Policy and Governance in Urban Sustainability

Governments and local authorities play a crucial role in driving the transition to sustainable urban development. Effective policy frameworks, regulations, and incentives are essential to promote sustainable practices in urban areas. For example, policies that encourage the use of renewable energy, the construction of energy-efficient buildings, and the protection of green spaces can significantly influence urban sustainability (Barlett, 2011). At the same time, urban governance must be inclusive, ensuring that all citizens have a voice in decision-making processes (Satterthwaite, 2016). Public participation in urban planning can lead to more equitable, effective, and socially inclusive solutions to environmental challenges. Furthermore, cities must invest in capacity-building for local governments to better manage sustainability efforts and collaborate across sectors (Müller, 2018). National and international frameworks, such as the United Nations Sustainable Development Goals (SDGs), also provide guidance and benchmarks for cities to follow in their sustainability efforts (UN, 2015).

Integrating Green Spaces into Urban Design

The integration of green spaces into urban design is a central aspect of urban sustainability. Urban green spaces, such as parks, gardens, and recreational areas, not only enhance the aesthetic value of cities but also provide numerous environmental, social, and health benefits (Tzoulas et al., 2007). These spaces act as carbon sinks, absorbing carbon dioxide and helping mitigate climate change (Bowler et al., 2010). They also provide crucial habitat for urban wildlife and contribute to biodiversity conservation. Additionally, green spaces improve air quality by filtering pollutants, reduce the urban heat island effect, and provide opportunities for physical activity and mental well-being (Kabisch et al., 2017). By increasing the number of green spaces in urban areas, cities can become more resilient to climate change and improve the overall quality of life for residents. Urban planners and architects must prioritize the inclusion of green infrastructure to create sustainable cities that promote ecological balance and social well-being.

Urban Heat Island Effect and Mitigation Strategies

The urban heat island (UHI) effect refers to the phenomenon where urban areas experience significantly higher temperatures than surrounding rural areas due to human activities, such as construction, transportation, and industrial processes (Oke, 1982).

The UHI effect exacerbates energy consumption, increases air pollution, and poses health risks to urban residents, particularly during heatwaves (Santamouris, 2015). Mitigating the UHI effect is critical for enhancing urban sustainability. One of the most effective strategies is increasing the presence of vegetation in cities, such as through the creation of green roofs, urban forests, and parks (Bowler et al., 2010). These green spaces provide shade and cool the air through evapotranspiration, reducing the ambient temperature (Liu et al., 2016). In addition, the use of reflective or cool roofing materials and the installation of permeable surfaces can help reduce the absorption of heat by buildings and roads, further mitigating the UHI effect (Akbari, 2005). By addressing the UHI effect, cities can become more comfortable, energy-efficient, and resilient to climate change.

Circular Economy in Urban Areas

The concept of a circular economy is gaining traction in urban sustainability discussions as an alternative to the traditional linear economy, which is based on a "take, make, dispose" model. In a circular economy, resources are used more efficiently, and waste is minimized through strategies such as recycling, reusing, and remanufacturing (Geissdoerfer et al., 2017). This approach aligns with the principles of sustainable urban development by reducing resource consumption, lowering greenhouse gas emissions, and promoting waste reduction. In urban areas, implementing a circular economy involves redesigning products and systems to minimize waste and ensure that materials can be reused or recycled (Lieder & Rashid, 2016). For example, cities can develop infrastructure for collecting and processing recyclables, incentivize businesses to use recycled materials, and promote the use of durable, repairable products. Additionally, waste-to-energy technologies can help cities convert non-recyclable waste into valuable resources, such as energy and fertilizers (Zhang et al., 2017). Adopting circular economy principles in urban areas is essential for reducing environmental impacts and creating more sustainable, resilient cities.

Green Building Practices and Sustainable Construction

The construction industry is a significant contributor to global environmental impacts, from resource depletion to carbon emissions. Green building practices aim to minimize these impacts by incorporating sustainable design principles and eco-friendly materials into the construction and operation of buildings (Kibert, 2016). One of the most effective ways to achieve sustainability in building construction is through energy-efficient design, which reduces the energy consumption of buildings over their lifespan (Yudelson, 2013). This includes optimizing insulation, utilizing natural lighting, and installing energy-efficient appliances. Moreover, the use of sustainable building materials, such as recycled materials, low-impact products, and locally sourced materials, reduces the environmental footprint of construction projects (Kibert, 2016). Green buildings also incorporate water-saving features, such as rainwater harvesting systems and low-flow fixtures, to reduce water consumption. Finally, green certification systems, such as LEED (Leadership in Energy and Environmental Design), provide a framework for evaluating and promoting sustainable building practices in urban areas (Yudelson, 2013). By adopting green building practices, cities can create energy-efficient, environmentally friendly structures that contribute to urban sustainability.

Urban Mobility and Sustainable Transport Solutions

Sustainable urban mobility is essential for reducing the environmental impact of cities, particularly concerning air pollution, greenhouse gas emissions, and traffic congestion. Transitioning to more sustainable transportation options, such as public transit, cycling, and walking, is a key component of urban sustainability. Public transportation systems that rely on clean energy, such as electric buses and trains, help reduce the dependence on fossil fuels and lower emissions (Creutzig et al., 2015). Additionally, promoting cycling and walking not only helps reduce traffic congestion but also improves public health by encouraging physical activity (Pucher et al., 2010). Cities can further enhance sustainable mobility by implementing bike-sharing programs, pedestrian-friendly infrastructure, and electric vehicle (EV) charging stations. Carpooling, ridesharing, and the development of low-emission zones also contribute to reducing urban air pollution and carbon footprints (Creutzig et al., 2015). Moreover, integrating smart technology, such as traffic management systems and mobility apps, can improve the efficiency of urban transportation networks. Ultimately, adopting sustainable transport solutions can make cities more livable, reduce their environmental impact, and enhance quality of life.

The Role of Technology in Sustainable Urban Development

Technology plays a pivotal role in advancing sustainable urban development by enabling smarter, more efficient systems that reduce environmental impacts and improve the quality of life for city residents. The integration of information and communication technologies (ICT) into urban infrastructure, commonly referred to as "smart cities," allows for better resource management, improved service delivery, and enhanced citizen engagement (Harrison et al., 2010). For example, smart grids enable the efficient distribution of energy, while sensors and data analytics help optimize waste collection and water usage (Hashem et al., 2016). Additionally, the use of advanced building technologies, such as energy-efficient appliances, solar panels, and automated energy management systems, contributes to lowering energy consumption and reducing greenhouse gas emissions (Bibri & Krogstie, 2017). Digital platforms can also facilitate more sustainable mobility by promoting car-sharing and on-demand transit services. Furthermore, technology supports urban resilience by enhancing climate forecasting, improving disaster preparedness, and enabling the restoration of ecosystems. As cities become increasingly digital, the effective use of technology is critical to achieving long-term sustainability goals.

Environmental Justice and Sustainable Urban Development

Environmental justice is a critical aspect of urban sustainability, as it ensures that all individuals, regardless of their socio-economic status, race, or ethnicity, have equal access to a healthy environment. In many cities, marginalized communities often bear the brunt of environmental degradation, including air pollution, inadequate waste management, and lack of access to green spaces (Sze, 2006). Addressing these disparities is essential for creating truly sustainable cities. Urban planners must prioritize equitable development by ensuring that vulnerable populations have access to resources such as clean air, water, and affordable housing. Policies aimed at reducing environmental hazards, such as stricter pollution controls and the expansion of public transportation, can help address these inequities (Bullard, 2007). Furthermore, community participation in decision-making processes ensures that all voices are heard and that sustainability initiatives benefit everyone (Foster, 2011). Achieving environmental justice not only improves the well-being of marginalized communities but also contributes to the overall sustainability of cities by promoting inclusivity and reducing social inequalities.

Reducing Urban Waste Through Circular Economy Practices

Urban areas are responsible for a significant portion of global waste production. To address this issue, many cities are turning to circular economy practices, which aim to reduce waste through recycling, reusing, and rethinking the life cycle of products. In a circular economy, waste is seen as a resource rather than something to be discarded (Ellen MacArthur Foundation, 2013). Cities can adopt policies and technologies that promote waste separation, composting, and the repurposing of materials to create a more sustainable and efficient urban waste system (Geissdoerfer et al., 2017). For example, cities can set up dedicated recycling bins for paper, glass, and plastics, while encouraging businesses to reduce packaging and invest in recyclable or biodegradable materials. Moreover, urban infrastructure, such as waste-to-energy plants, can help divert non-recyclable waste from landfills and convert it into usable resources, such as electricity and heating (Zhang et al., 2017). By embracing circular economy principles, cities can significantly reduce their environmental footprint, decrease reliance on landfills, and contribute to global sustainability goals.

The Importance of Community Engagement in Urban Sustainability

Community engagement is a critical aspect of achieving urban sustainability. Involving residents in decision-making processes ensures that sustainability initiatives reflect the needs and preferences of the community, fostering a sense of ownership and responsibility (Choguill, 2014). Community engagement can take many forms, from participatory planning processes to local sustainability initiatives, such as tree planting programs or waste reduction campaigns (Friedmann, 1992). By actively involving citizens in sustainability efforts, cities can harness local knowledge, build trust, and ensure that policies and projects are both effective and equitable (Arnstein, 1969). Moreover, community involvement fosters collaboration between diverse stakeholders, such as local governments, businesses, and non-profit organizations, which can lead to innovative solutions to urban challenges. Ultimately, the success of urban sustainability initiatives depends on the active participation and engagement of residents, as they play a central role in shaping the future of their cities.

Conclusion: Building Sustainable Cities for the Future

In conclusion, the creation of sustainable cities is a critical challenge in the face of rapid urbanization, climate change, and growing environmental concerns. Sustainable urban development requires an integrated approach that incorporates policies, technologies, infrastructure, and community engagement to promote environmental stewardship, economic resilience, and social equity (United Nations, 2015). By embracing renewable energy, green building practices, circular economy principles, and sustainable transportation solutions, cities can reduce their environmental impact while enhancing the well-being of their residents (Satterthwaite, 2016). Furthermore, the involvement of citizens in sustainability efforts ensures that urban development meets the needs of all stakeholders and fosters inclusive, sustainable communities. Collaborative efforts between governments, businesses, and individuals are essential to create urban spaces that are resilient to climate change and able to provide a high quality of life for future generations (Satterthwaite, 2016). The transition to sustainable urban environments is not only a matter of reducing environmental footprints but also of fostering social inclusivity and economic opportunities for all city residents. As the global population increasingly shifts toward urban areas, it is imperative that cities adopt sustainable practices that protect natural resources, enhance public health, and promote economic prosperity. Ultimately, sustainable cities must be viewed as dynamic systems, where environmental, social, and economic factors are inextricably linked, ensuring a balanced approach to urban development. The future of cities depends on our ability to act now to create green, resilient, and equitable urban spaces.

Chapter 11: Environmental Justice: Addressing inequalities

Introduction to Environmental Justice

Environmental justice refers to the fair treatment and meaningful involvement of all people, regardless of their race, color, national origin, or income, with respect to environmental laws, regulations, and policies (United States Environmental Protection Agency [EPA], 2021). Historically, marginalized communities have borne a disproportionate share of environmental harms, including pollution, hazardous waste, and inadequate access to resources like clean air and water. These communities often face multiple barriers, such as poverty, lack of political power, and systemic discrimination, that hinder their ability to advocate for environmental protection. Addressing environmental justice is not only about remedying these disparities but also about creating equitable opportunities for sustainable development. Achieving environmental justice involves recognizing and dismantling the structural inequalities that lead to environmental burdens in these communities and ensuring that all individuals have the opportunity to live in a safe and healthy environment (Bullard, 2007). This paper will explore the concept of environmental justice, the history of environmental inequalities, and contemporary efforts to address these disparities in communities around the world.

Historical Context of Environmental Inequities

The roots of environmental injustice can be traced back to historical patterns of segregation, colonization, and industrial development. In the United States, for instance, communities of color and low-income neighborhoods have long been the target sites for the placement of industrial facilities, landfills, and hazardous waste dumps (Cole & Foster, 2001). These practices have led to a legacy of pollution in communities already burdened by economic deprivation and limited access to healthcare. The Civil Rights Movement in the 1960s and 1970s provided a platform for advocating against such inequities, leading to the eventual emergence of the environmental justice movement (Bullard, 1990). The 1982 protests in Warren County, North Carolina, against the siting of a toxic waste landfill in a predominantly Black community are often cited as a pivotal moment in the movement. This event helped bring national attention to the intersection of environmental degradation and racial discrimination, setting the stage for future legal and policy reforms aimed at addressing environmental justice issues.

The Environmental Justice Movement

The environmental justice movement gained significant momentum in the 1980s and 1990s, following widespread grassroots activism and community mobilization. As communities of color and low-income populations increasingly recognized the disproportionate environmental burdens they were facing, they began organizing to demand cleaner environments and fairer treatment from government and industry (Gottfried & Pellow, 2009). One of the key aspects of the environmental justice movement is its focus on the intersectionality of environmental issues and social justice. Environmental justice advocates argue that pollution, climate change, and other environmental problems are not just issues of ecology but also of human rights and equality. Over time, the movement has succeeded in raising awareness of the importance of considering race, class, and other factors in environmental policies and decision-making (Mohai, Pellow, & Roberts, 2009). In response, there have been efforts at the local, national, and international levels to integrate environmental justice principles into law and governance. These efforts have included the establishment of environmental justice policies, such as the U.S.

Environmental Protection Agency's (EPA) Environmental Justice Strategy, which aims to address the disproportionate environmental impacts on vulnerable communities.

Key Principles of Environmental Justice

At the heart of environmental justice are several key principles that seek to ensure that no group bears a disproportionate burden of environmental harm. The first principle is that all people have the right to a healthy environment, regardless of their background, race, or income (United Nations, 2015). This principle emphasizes the need to recognize the inequitable distribution of environmental risks and to rectify these disparities. Another important principle is the recognition of the right to participate in decision-making processes regarding environmental policies and actions that affect communities. This includes involving marginalized populations in discussions about land use, zoning laws, and public health policies (Agyeman, Bullard, & Evans, 2003). A third principle is the need for governments and industries to take proactive measures to prevent environmental harm before it occurs, rather than simply responding to problems once they arise. The ultimate goal of these principles is to foster a more equitable, inclusive, and sustainable approach to environmental governance.

Environmental Injustice and Health Disparities

One of the most pressing concerns in the realm of environmental justice is the connection between environmental degradation and public health. Communities that are exposed to higher levels of pollution and toxic waste often suffer from a range of health issues, including respiratory diseases, cardiovascular problems, and cancer. These health disparities are exacerbated in low-income and minority communities, where access to healthcare is limited and social determinants of health are already poor (Krieger, 2001). For example, studies have shown that people living near industrial sites or waste disposal facilities experience higher rates of asthma, lead poisoning, and other chronic conditions (Gould & Schnaiberg, 2006). Moreover, climate change is expected to exacerbate these health inequalities by increasing the frequency of extreme weather events, such as heatwaves and floods, which disproportionately affect vulnerable populations. Addressing environmental injustice, therefore, requires a comprehensive approach that includes both environmental protections and public health initiatives to mitigate these adverse health outcomes.

Environmental Justice and Climate Change

Climate change is one of the most significant environmental justice issues of the 21st century. Vulnerable communities are disproportionately affected by the impacts of climate change, including rising sea levels, extreme weather events, and changing agricultural patterns (Schlosberg & Collins, 2014). For instance, low-income populations and people of color are more likely to live in flood-prone areas or regions that are subject to extreme heat, both of which are exacerbated by climate change. Furthermore, these communities often have fewer resources to adapt to these changes, making them more vulnerable to the economic and health impacts of climate change (Hernandez, 2012). Addressing climate justice involves recognizing these inequities and ensuring that vulnerable communities have access to the resources, support, and infrastructure necessary to cope with climate impacts. This includes investing in climate resilience, such as flood defenses and heat mitigation strategies, in areas that are most at risk. It also requires shifting from a focus on mitigating climate change to also include strategies for adapting to its inevitable impacts.

The Role of Policy in Environmental Justice

Policy plays a crucial role in advancing environmental justice by addressing the systemic inequities that disproportionately affect marginalized communities. Legal frameworks, such as the National Environmental Policy Act (NEPA) and the Clean Air Act in the

United States, are vital tools for ensuring that government decisions consider the potential environmental and health impacts on vulnerable populations (Bernard, 2017). In addition to regulatory policies, there has been increasing advocacy for the inclusion of environmental justice principles in environmental assessments and planning processes. These efforts have led to the development of guidelines that require public participation, particularly from communities that are most affected by environmental policies. Furthermore, international frameworks, such as the United Nations' Sustainable Development Goals (SDGs), emphasize the importance of environmental justice and the need to reduce inequality within and among countries. Moving forward, the effectiveness of environmental justice policies will depend on strong enforcement, sufficient funding, and ongoing community engagement to ensure that vulnerable populations are protected and empowered.

Grassroots Movements and Environmental Justice Advocacy

Grassroots movements have been pivotal in driving the environmental justice agenda forward. Communities that face disproportionate environmental harms often lack the political and economic power to influence decision-making processes. As a result, local community-based organizations and advocacy groups play a crucial role in raising awareness, mobilizing support, and demanding policy changes that address environmental injustices. For example, the environmental justice movement in the United States was significantly shaped by the work of activists like Dr. Robert Bullard, who has been instrumental in highlighting the links between race, class, and environmental degradation (Bullard, 2005). These grassroots movements have also been successful in pushing for the inclusion of environmental justice principles in environmental policy frameworks. In addition, they advocate for community-driven solutions to environmental problems, such as the relocation of toxic waste facilities, stricter pollution regulations, and the promotion of sustainable urban planning. By empowering communities to take ownership of their environmental future, these movements contribute to a more equitable and inclusive approach to environmental governance (Checker, 2011).

Environmental Justice in Global Context

While environmental justice has often been discussed within the context of the United States, it is a global issue that affects many parts of the world. In countries in the Global South, communities that are economically disadvantaged and marginalized also experience the harmful effects of environmental degradation, often as a result of both historical and contemporary forms of colonialism and exploitation. For instance, large multinational corporations often exploit natural resources in low-income countries, leaving behind environmental destruction and health hazards for local populations (Kuehn, 2017). Countries in Africa, Asia, and Latin America have seen rapid industrialization without the infrastructure and regulations needed to protect vulnerable populations. Furthermore, the negative effects of climate change, such as droughts and floods, are disproportionately felt in these regions. Environmental justice on a global scale involves ensuring that these communities have a voice in international environmental governance and that global environmental policies take into account the needs and rights of vulnerable populations. As climate change continues to have a global impact, there is a growing recognition of the need for a more inclusive approach to environmental policy that addresses both local and global inequities (Pellow, 2018).

Indigenous Communities and Environmental Justice

Indigenous communities are often at the forefront of the struggle for environmental justice, as their ancestral lands and way of life are directly impacted by environmental degradation. These communities often possess unique cultural knowledge about the land and its ecosystems, making them valuable allies in conservation efforts. However, they also face the added burden of having their rights and traditions ignored or violated by governments and industries seeking to exploit natural resources (Mann, 2018). From deforestation in the Amazon to oil extraction in the Arctic, Indigenous communities are often the first to witness and resist the environmental harms caused by industrial development. Environmental justice for Indigenous peoples involves recognizing their sovereignty, protecting their land rights, and ensuring that they are meaningfully involved in decision-making processes related to the environment. Many Indigenous groups advocate for the preservation of traditional ecological knowledge and sustainable resource management practices that have been passed down through generations. By amplifying the voices of Indigenous communities, the environmental justice movement can address not only environmental harms but also social and cultural inequities (Corntassel, 2012).

The Role of Education in Advancing Environmental Justice

Education plays a pivotal role in advancing the goals of environmental justice. By raising awareness about the systemic nature of environmental inequalities and the intersectionality between environmental and social justice, education can empower communities to advocate for their rights and protect their environment. Schools, universities, and community organizations can serve as platforms for environmental justice education, providing both theoretical knowledge and practical tools for addressing local environmental issues. Environmental justice curricula often incorporate the history of environmental racism, the science of environmental health, and the social and economic dimensions of sustainability. Furthermore, educating the next generation of environmental leaders helps to ensure that the principles of justice, equity, and inclusivity are embedded in future policy and governance frameworks. At the same time, community-based education programs can help to bridge the gap between scientific research and local knowledge, fostering collaboration between activists, scientists, and policymakers (Mueller, 2017). In this way, education is both a tool for raising consciousness and a means of creating tangible solutions to environmental inequalities.

Corporate Responsibility and Environmental Justice

Corporate responsibility plays a critical role in addressing environmental justice, as many of the most significant environmental harms are caused by industrial activities and corporate practices. Corporations, particularly in the fossil fuel, mining, and manufacturing industries, have been responsible for environmental degradation that disproportionately affects vulnerable communities. Environmental justice advocates call for greater corporate accountability in addressing environmental harms, ensuring that companies not only comply with environmental regulations but also actively work to prevent harm in the communities where they operate (Sullivan, 2013). This involves adopting cleaner technologies, improving waste management practices, and engaging with local communities to understand their concerns and needs. Corporate social responsibility (CSR) initiatives are increasingly being used as tools for mitigating the environmental and social impacts of corporate actions. However, critics argue that CSR is often insufficient and that more stringent regulations and enforcement are needed to ensure that corporations genuinely prioritize the well-being of marginalized populations (Bakan, 2004). By holding corporations accountable for their environmental and social impacts, society can promote a more just and sustainable future for all.

The Role of the Government in Environmental Justice

Governments play a central role in the advancement of environmental justice by creating and enforcing laws that protect marginalized communities from environmental harms. Environmental regulations, such as the Clean Water Act and the Clean Air Act, have historically helped to address some of the environmental burdens faced by vulnerable populations (Gottfried & Pellow, 2009). However, environmental justice advocates argue that these laws have often been insufficient in addressing the disproportionate impacts of pollution on low-income communities and communities of color. In recent years, governments have increasingly recognized the need for more targeted policies to address environmental justice, such as the implementation of environmental justice screening tools and the inclusion of vulnerable communities in environmental decision-making processes (Mohai et al., 2011). These efforts reflect a growing recognition that environmental harm is not distributed equally, and that a more equitable approach to environmental policy is necessary. Government action must go beyond reactive measures and prioritize the prevention of harm through proactive environmental policies and the protection of vulnerable populations.

International Efforts and Environmental Justice

International efforts to promote environmental justice have gained traction in recent years, especially as environmental degradation and climate change have become increasingly global issues. International organizations, such as the United Nations (UN) and the World Health Organization (WHO), have increasingly recognized the importance of environmental justice in achieving global sustainability goals. For example, the UN's 2030 Agenda for Sustainable Development explicitly includes the concept of "leaving no one behind," which emphasizes the need to ensure that all people, including marginalized communities, benefit from environmental policies (United Nations, 2015). Similarly, the Paris Agreement on climate change acknowledges the need for a just transition to a low-carbon economy, taking into account the needs of vulnerable populations (UNFCCC, 2015). These international frameworks offer an opportunity to address environmental injustice on a global scale, ensuring that the voices of marginalized communities are included in discussions about climate policy, development, and environmental governance. By linking environmental justice to global sustainability, international organizations can help facilitate a more equitable approach to environmental challenges.

Environmental Justice and the Future

The future of environmental justice lies in the continued recognition of the interconnectedness between environmental and social inequalities. As the world faces unprecedented environmental challenges, such as climate change, resource depletion, and biodiversity loss, it is critical that these challenges are addressed in a way that prioritizes the needs of vulnerable communities. Moving forward, environmental justice must be integrated into all aspects of environmental governance, from local policies to global climate agreements. This will require continued collaboration between governments, businesses, and civil society organizations, as well as greater investment in research and education to develop sustainable and just solutions. Moreover, it will require a shift in how we think about environmental problems, moving beyond the idea that environmental degradation is an isolated issue to recognizing that it is deeply connected to broader issues of poverty, inequality, and human rights. By centering justice in our environmental policies and actions, we can create a more equitable and sustainable future for all.

Integrating Environmental Justice into Policy Development

Environmental justice must be incorporated into every stage of policy development to ensure its effectiveness in addressing the needs of marginalized communities. This means that policies should be designed to not only reduce environmental harm but also to promote social equity. Policymakers can take several approaches to achieve this, such as incorporating environmental justice impact assessments, consulting with affected communities, and creating stronger legal frameworks for environmental protections. Additionally, ensuring that marginalized groups have access to the necessary resources to participate in policymaking—such as education, legal support, and financial assistance—is critical to achieving meaningful change (Faber & O'Connor, 2012). By integrating environmental justice into policy, governments can create policies that are both equitable and effective in promoting environmental sustainability. The role of non-governmental organizations (NGOs) and grassroots groups in influencing policy changes is also critical, as these groups provide a voice for those who are often excluded from decision-making processes (Holifield, 2009).

Addressing Disproportionate Health Impacts of Environmental Injustice

One of the most significant aspects of environmental justice is its connection to public health. Vulnerable communities often bear the brunt of pollution, toxic waste, and other environmental hazards that lead to a range of health problems, including respiratory issues, cancers, and waterborne diseases. Research has shown that low-income and minority communities are disproportionately affected by environmental pollutants such as industrial emissions, pesticides, and contaminated water sources (Bullard, 2000). These environmental health disparities contribute to higher rates of chronic illnesses in these communities, further exacerbating existing inequalities. Addressing environmental justice requires a multi-faceted approach to healthcare, including stricter environmental regulations, greater access to healthcare services, and improved public health interventions in affected communities. By focusing on reducing environmental health risks and improving health outcomes in vulnerable populations, we can work toward creating a more equitable society where everyone has access to a safe and healthy environment (Lerner, 2010).

The Role of Corporations in Environmental Justice

While governments have a key role in addressing environmental justice, corporations must also be held accountable for their actions. Corporations often cause or exacerbate environmental harm, particularly in marginalized communities where lax environmental regulations and oversight may allow industrial practices to persist unchecked. These industries are responsible for much of the pollution, waste, and resource exploitation that contribute to environmental justice issues (Sullivan, 2013). However, businesses also have the potential to be part of the solution. Corporate social responsibility (CSR) initiatives can support environmental justice efforts by prioritizing sustainable practices, reducing emissions, and investing in community-led environmental projects. Multinational corporations that operate in multiple countries should ensure that their operations adhere to international environmental standards, ensuring that they do not exacerbate existing environmental inequalities. By adopting environmentally responsible practices and engaging in transparent dialogue with affected communities, businesses can help mitigate the environmental burdens placed on vulnerable populations (Bakan, 2004).

The Intersection of Climate Change and Environmental Justice

Climate change has become one of the most pressing environmental issues of our time, and it intersects with the goals of environmental justice in profound ways. The effects of climate change are not felt equally across the world—low-income communities and communities of color often face the most severe consequences. These communities tend to live in areas that are more vulnerable to extreme weather events, such as floods, hurricanes, and heatwaves. Furthermore, they have fewer resources to recover from such disasters, making them more susceptible to the long-term impacts of climate change (Mohai et al., 2011). The environmental justice movement highlights the need for policies that not only reduce greenhouse gas emissions but also ensure that vulnerable communities are equipped to adapt to climate change. A just transition to a sustainable and low-carbon future requires prioritizing the needs of these communities, ensuring that they are not left behind in the transition to a greener economy. This means investing in renewable energy, sustainable infrastructure, and climate resilience strategies in disadvantaged areas (Agyeman et al., 2016).

The Future of Environmental Justice and Global Collaboration

Looking ahead, the future of environmental justice will depend on continued collaboration at local, national, and global levels. As the effects of environmental degradation and climate change continue to intensify, it will be essential for all stakeholders—including governments, corporations, communities, and international organizations—to come together to create equitable solutions. Environmental justice must be recognized as a fundamental human right, and policies must reflect this commitment by ensuring that all people, regardless of race, class, or nationality, have access to a clean and healthy environment. In this globalized world, environmental justice cannot be achieved in isolation; it requires a collective effort to address systemic inequalities and create a more just and sustainable future for all. By building on the progress made in environmental justice movements and ensuring that the voices of marginalized communities continue to be heard, we can achieve a more inclusive approach to environmental governance (Pellow, 2018).

Chapter 12: Green Technology: Innovations for the future

Introduction to Green Technology

Green technology, also known as environmental technology, refers to innovations and practices that minimize harm to the environment while improving sustainability. As the world faces growing concerns regarding climate change, pollution, and resource depletion, green technology plays an increasingly pivotal role in offering solutions for a more sustainable future. Green technologies aim to reduce carbon footprints, conserve natural resources, and increase energy efficiency across various sectors, including energy, transportation, manufacturing, and waste management (Alsema, 2006). Innovations such as renewable energy systems, electric vehicles, and sustainable agriculture practices are prime examples of how technology can be harnessed to address environmental challenges. The shift toward green technology is essential not only for mitigating the effects of climate change but also for creating new economic opportunities in industries related to clean energy, recycling, and environmental conservation (Morrison et al., 2020).

As the world transitions into an era of cleaner technologies, the importance of research and development in green technology cannot be overstated. Governments, corporations, and academic institutions must collaborate to foster innovation and implement policies that support green technology adoption at scale. Green technologies are expected to play a crucial role in both mitigating environmental harm and fostering a more resilient and sustainable global economy.

Renewable Energy as a Key Green Technology

Renewable energy is one of the cornerstones of green technology, offering a sustainable alternative to fossil fuels. Solar, wind, hydroelectric, and geothermal power are at the forefront of the transition to green energy systems. Solar energy, for instance, harnesses sunlight using photovoltaic cells, while wind energy converts wind flow into electrical power using turbines. Both methods significantly reduce reliance on non-renewable sources such as coal and oil, which contribute to greenhouse gas emissions (Kaldellis et al., 2012). The development of renewable energy systems has been accelerated by advances in materials science, energy storage technologies, and grid integration. Solar panels, for example, have become more efficient and cost-effective, making them a viable option for both residential and industrial applications (Wang et al., 2020).

Wind power has similarly grown in capacity, with offshore wind farms now contributing a significant share of energy generation in countries like Denmark and the United Kingdom. Hydroelectric and geothermal power also play important roles in renewable energy generation. Together, these technologies offer a robust solution to meet growing global energy demands while mitigating the environmental impacts associated with fossil fuel consumption. Investment in renewable energy infrastructure is critical for global efforts to combat climate change and reduce environmental degradation.

Energy-Efficient Buildings: A Green Technology Solution

Energy-efficient buildings represent one of the most immediate ways to reduce energy consumption and environmental impact. By incorporating green technologies into the design, construction, and operation of buildings, it is possible to achieve significant reductions in energy use, water consumption, and carbon emissions. Key innovations include the use of energy-efficient insulation, smart lighting, advanced heating and cooling systems, and green roofs. These technologies not only reduce environmental impact but also contribute to cost savings in the long run by decreasing energy bills (Pérez-Lombard et al., 2008).

Building codes and standards that promote energy efficiency have become increasingly important in encouraging sustainable construction. In particular, the LEED (Leadership in Energy and Environmental Design) certification has been widely adopted to recognize buildings that meet high environmental performance standards. Energy-efficient buildings also support better indoor air quality, improved occupant health, and higher levels of productivity. Moreover, sustainable building practices can extend beyond energy use and incorporate the use of environmentally friendly materials, rainwater harvesting systems, and waste management technologies. With the global urban population on the rise, the integration of energy-efficient technologies into the construction of both residential and commercial buildings is a vital component of sustainable urban development (González et al., 2012).

Electric Vehicles: The Future of Sustainable Transportation

The transportation sector is a major contributor to global greenhouse gas emissions, with traditional gasoline and diesel-powered vehicles being the primary source of air pollution and carbon emissions (Hawkins et al., 2013). Electric vehicles (EVs) represent one of the most promising green technology innovations for transforming the transportation sector. By using electricity instead of fossil fuels, EVs have the potential to reduce air pollution, decrease reliance on oil, and significantly lower carbon footprints. In recent years, advancements in battery technology have led to longer driving ranges and faster charging times for EVs, making them increasingly attractive to consumers (Egbue & Long, 2012).

Governments worldwide are providing incentives for the adoption of electric vehicles, including tax credits, rebates, and subsidies for both consumers and manufacturers. These incentives, coupled with the development of charging infrastructure, are expected to drive the growth of the electric vehicle market in the coming years. Moreover, the integration of renewable energy sources into the electricity grid will make EVs even more sustainable by reducing their overall carbon impact. As electric vehicles become more mainstream, they will play a critical role in reducing emissions from the transportation sector and contributing to broader sustainability goals.

Smart Grids: Enhancing Energy Efficiency with Green Technology

Smart grids represent a transformative advancement in energy infrastructure that enhances the efficiency, reliability, and sustainability of power systems. A smart grid is an electricity network that uses digital technology to detect and respond to changes in electricity usage, ensuring that power is distributed efficiently and in real-time. By integrating renewable energy sources, such as solar and wind, into the grid, smart grids can better manage fluctuations in supply and demand. Additionally, smart meters allow consumers to track and optimize their energy use, leading to reduced energy consumption and lower utility bills (Tushar et al., 2014). The ability of smart grids to incorporate decentralized power generation, such as rooftop solar panels, helps reduce transmission losses and contributes to a more resilient energy system.

One of the main benefits of smart grids is their ability to reduce carbon emissions by optimizing the use of renewable energy. Through demand-response technologies, smart grids can automatically adjust power usage during peak hours, preventing overreliance on fossil fuel-based power plants. As more cities and regions adopt smart grid technologies, the potential to improve energy efficiency and reduce environmental impact grows exponentially. The global expansion of smart grids is vital to advancing green technology and ensuring that the shift to renewable energy systems is both efficient and sustainable (Georgilakis, 2015).

Sustainable Agriculture Technologies for a Green Future

Sustainable agriculture technologies are essential to feeding the world's growing population while protecting the environment. Traditional farming methods often lead to soil depletion, excessive water usage, and increased greenhouse gas emissions. Green technologies in agriculture aim to minimize these impacts while improving productivity. Precision farming techniques, which use data from sensors and satellites, enable farmers to apply water, fertilizers, and pesticides more efficiently, thus reducing waste and enhancing crop yields (Lal, 2007). Additionally, the use of organic farming practices and agroecological systems supports biodiversity, reduces dependency on chemical inputs, and promotes soil health.

Another important green technology in agriculture is vertical farming, which uses controlled indoor environments to grow crops in stacked layers, often using hydroponics or aeroponics. Vertical farming uses significantly less land and water compared to traditional farming and can be practiced in urban settings, reducing the need for long-distance food transportation. Furthermore, biotechnological advancements, such as genetically modified crops, have the potential to increase resistance to pests and diseases, further reducing the environmental footprint of agricultural production (Tilman et al., 2002). By integrating green technologies into agricultural practices, farmers can reduce their environmental impact while increasing food security in a more sustainable manner.

Green Building Materials: Innovations for Sustainable Construction

Green building materials are key components of sustainable construction practices. These materials are designed to reduce environmental impact during both production and use. For example, low-carbon cement and recycled steel can significantly reduce the carbon footprint of construction projects. Additionally, the use of renewable resources, such as bamboo and reclaimed wood, can help reduce the reliance on virgin materials and minimize deforestation (Kibert, 2016). Another example of a green building material is insulation made from natural or recycled materials, which helps improve energy efficiency by reducing the need for heating and cooling.

The choice of building materials is crucial not only for reducing emissions but also for improving the long-term sustainability of buildings. For instance, green roofs, which are covered with vegetation, provide natural insulation, absorb rainwater, and promote biodiversity. Additionally, energy-efficient windows and doors reduce heat loss and improve building performance. As awareness of the environmental impacts of construction grows, there is an increasing demand for sustainable materials in both residential and commercial building projects. Green building certifications, such as LEED, help guide the selection of materials and ensure that buildings meet high sustainability standards (Jones et al., 2013).

The Role of Artificial Intelligence in Green Technology

Artificial intelligence (AI) is revolutionizing the field of green technology by providing innovative solutions to optimize energy use, reduce waste, and improve sustainability. AI-powered systems can analyze vast amounts of data from sensors, satellites, and smart meters to optimize energy distribution, predict maintenance needs for renewable energy systems, and enhance efficiency in manufacturing processes (Li et al., 2020). In the energy sector, AI algorithms can predict energy demand patterns and help integrate renewable energy sources, such as wind and solar, into the grid more efficiently.

In agriculture, AI is used for precision farming, helping farmers make data-driven decisions about irrigation, pest control, and crop management. AI-powered drones and robots can monitor crop health, detect diseases, and automate harvesting, reducing the need for chemical pesticides and minimizing environmental impact. Furthermore, AI has applications in waste management by improving recycling rates and optimizing waste collection schedules. As AI continues to evolve, its potential to contribute to the green technology sector grows, offering solutions that can mitigate environmental challenges while improving overall sustainability (Vinuesa et al., 2020).

Green Transportation Infrastructure: Supporting Sustainable Mobility

Green transportation infrastructure plays a vital role in promoting sustainable mobility solutions. As cities continue to grow, the need for efficient, eco-friendly transport systems has become more urgent. Public transportation systems, including buses, trains, and subways, are central to reducing urban carbon emissions and traffic congestion. By using renewable energy to power public transit networks and integrating energy-efficient vehicles, cities can significantly reduce their environmental impact (Sovacool, 2017). Electric buses, for example, have emerged as a promising solution to reduce air pollution in urban centers.

Moreover, the expansion of cycling and pedestrian infrastructure supports sustainable transportation by encouraging individuals to use non-motorized modes of transport. Cities like Copenhagen and Amsterdam have invested heavily in bike lanes and walkable urban environments, contributing to lower levels of pollution and improved public health. Additionally, the development of electric vehicle (EV) charging stations and smart parking systems encourages the adoption of electric cars, further reducing the carbon footprint of the transportation sector. By supporting green transportation infrastructure, cities can create more sustainable, livable environments while mitigating climate change impacts (Pucher et al., 2010).

The Role of Circular Economy in Green Technology

The circular economy (CE) is a critical concept in advancing green technologies and promoting sustainability. Unlike the traditional linear economy, which follows a "take, make, dispose" model, the circular economy focuses on extending the lifecycle of products by reusing, recycling, and refurbishing materials. Green technologies are essential for the successful implementation of a circular economy, as they provide the tools and systems necessary to manage resources efficiently and minimize waste (Geissdoerfer et al., 2017).

For instance, in the electronics sector, the use of eco-design principles allows products to be manufactured with the intention of easier disassembly, repair, and recycling. Moreover, innovative recycling technologies, such as e-waste recycling and plastic-to-fuel conversion, are crucial for reducing the environmental impact of discarded materials. The circular economy also emphasizes the importance of sustainable product design, resource recovery, and minimizing the extraction of raw materials from the environment. By adopting a circular economy approach, industries can significantly reduce waste, conserve natural resources, and contribute to a greener, more sustainable future (Murray et al., 2017).

Sustainable Water Technologies: Ensuring Clean Water for All

Access to clean water is one of the most critical challenges faced by communities worldwide. Sustainable water technologies are essential in addressing the growing demand for water while ensuring its responsible use. Innovations such as desalination plants, water filtration systems, and rainwater harvesting techniques help communities access clean water without overexploiting natural freshwater resources. Desalination, for instance, uses reverse osmosis or thermal processes to convert seawater into drinkable water, providing a viable solution for water-scarce regions (Elimelech & Phillip, 2011).

Water recycling technologies, such as greywater systems, are also playing an important role in reducing water consumption in both urban and agricultural settings. These systems filter and purify wastewater for non-potable uses like irrigation or industrial processes, significantly reducing the strain on freshwater sources. Additionally, smart water management systems that incorporate real-time data analytics help monitor water usage and detect leaks, ensuring that water resources are used efficiently and sustainably. By investing in and adopting these technologies, societies can protect this vital resource for future generations (Xie et al., 2020).

Waste-to-Energy Technologies: Converting Waste into Resources

Waste-to-energy (WTE) technologies are an innovative solution for managing waste and generating renewable energy simultaneously. These technologies involve the conversion of municipal solid waste, agricultural residues, or industrial waste into usable forms of energy, such as electricity or heat. Incineration, anaerobic digestion, and gasification are common methods used in WTE facilities (Zhou et al., 2016). These processes not only reduce the volume of waste that ends up in landfills but also help reduce the reliance on fossil fuels for energy production.

WTE technologies offer significant environmental benefits by decreasing methane emissions from landfills, which are a potent greenhouse gas. Moreover, the energy produced through WTE processes can be used to power homes, businesses, and even entire cities, contributing to renewable energy targets. However, it is crucial to ensure that WTE facilities are designed with appropriate pollution controls to minimize harmful emissions. In the long term, WTE technologies can help create a circular economy by turning waste into valuable resources, reducing both environmental pollution and the strain on landfill space (Moran et al., 2014).

Green Data Centers: Reducing the Environmental Impact of Technology

Data centers, which house the servers and infrastructure necessary to support digital technologies, are a growing source of energy consumption worldwide. As the demand for cloud computing and data storage increases, so does the need for energy-efficient and environmentally friendly solutions. Green data centers aim to reduce their carbon footprint by adopting energy-efficient designs, renewable energy sources, and innovative cooling techniques. For example, some data centers use ambient air cooling or liquid cooling to reduce the need for traditional air conditioning systems, which consume large amounts of electricity (Shehabi et al., 2016).

Additionally, the integration of renewable energy sources, such as solar and wind, can help power data centers without relying on fossil fuels. The use of virtualized servers and efficient hardware also allows data centers to optimize energy usage by reducing the number of physical machines required. As the digital world continues to grow, green data centers are essential in reducing the environmental impact of the technology sector. By transitioning to energy-efficient and sustainable practices, data centers can support the ongoing development of digital technologies while helping mitigate climate change (Jones et al., 2018).

Green Manufacturing: Reducing Industrial Emissions

Green manufacturing focuses on minimizing the environmental impact of industrial processes by improving efficiency, reducing waste, and utilizing sustainable materials. One key aspect of green manufacturing is the adoption of cleaner production technologies, which help reduce the amount of energy, water, and raw materials needed to produce goods (Tartakovsky et al., 2013). In addition, sustainable manufacturing practices aim to minimize emissions of harmful pollutants, such as greenhouse gases and particulate matter, by using cleaner energy sources and implementing pollution control technologies.

The circular economy plays a critical role in green manufacturing by encouraging the reuse and recycling of materials. For example, companies are increasingly designing products with the intention of facilitating easy disassembly and recycling at the end of their lifecycle. Furthermore, the use of renewable energy sources, such as solar and wind, in manufacturing processes helps reduce reliance on fossil fuels and lowers carbon emissions. By adopting green manufacturing practices, industries can contribute to environmental sustainability while also reducing costs and improving operational efficiency (Rossi et al., 2017).

The Future of Green Technology: Trends and Innovations

The future of green technology is promising, with innovations across various sectors offering solutions to some of the world's most pressing environmental challenges. Advancements in energy storage technologies, such as lithium-ion and solid-state batteries, are expected to facilitate the widespread adoption of renewable energy sources by addressing the intermittency issues associated with solar and wind power (Chen et al., 2017). In the transportation sector, hydrogen fuel cells and autonomous electric vehicles have the potential to revolutionize how goods and people are transported, further reducing emissions from the transportation industry.

Additionally, new technologies are being developed to capture and store carbon emissions, such as carbon capture and storage

(CCS) systems. These technologies could play a critical role in mitigating the effects of climate change by reducing the amount of CO2 released into the atmosphere. As governments, industries, and consumers continue to embrace sustainable practices, the growth of green technologies will be essential in achieving global environmental goals, including net-zero emissions and the preservation of natural ecosystems (Schneider et al., 2020).

Green Energy Solutions for Developing Countries

Green energy solutions are essential in helping developing countries transition towards more sustainable energy systems. Many developing nations face challenges in accessing reliable and affordable electricity, while also needing to mitigate environmental degradation caused by fossil fuel reliance. Renewable energy technologies, such as solar, wind, and hydropower, offer viable alternatives for meeting energy needs while reducing environmental impacts (Diaz et al., 2018). Solar energy, in particular, has proven to be an accessible and cost-effective solution in rural and off-grid areas, where traditional electricity infrastructure is lacking.

In addition to providing reliable power, renewable energy systems can also create jobs and foster economic development. For instance, the installation and maintenance of solar panels and wind turbines require skilled labor, which can lead to the growth of local industries. Moreover, green energy solutions can reduce the dependency on imported fossil fuels, improving energy security and contributing to the long-term sustainability of developing economies. By embracing renewable energy technologies, developing nations can achieve sustainable development goals while combating climate change and reducing their carbon footprint (Sovacool et al., 2019).

Green Building Innovations: Creating Sustainable Urban Spaces

Green building innovations have become an essential part of efforts to reduce the environmental impact of urban development. Green buildings utilize energy-efficient designs, sustainable materials, and advanced technologies to minimize their environmental footprint while providing comfortable and healthy living and working spaces. Key features of green buildings include energy-efficient heating and cooling systems, water-saving fixtures, and the use of renewable energy sources such as solar panels or wind turbines (Sartori & Hestnes, 2007).

Building certifications, such as LEED (Leadership in Energy and Environmental Design), have become industry standards for promoting green construction practices. These certifications incentivize developers to create sustainable buildings by rewarding efforts to reduce energy consumption, water usage, and waste production. Moreover, green buildings are designed with an emphasis on natural lighting and improved air quality, contributing to better overall well-being for occupants. As the world continues to urbanize, the adoption of green building practices is crucial in ensuring that cities remain livable, resource-efficient, and resilient to climate change (Kibert, 2016).

Green Manufacturing and the Role of Automation

Green manufacturing has been increasingly linked to the development and integration of automation technologies. These technologies are enabling manufacturers to optimize energy usage, reduce waste, and enhance the efficiency of production processes. Automation allows for greater precision in manufacturing, reducing material waste and improving the quality of products, all of which contribute to a greener production process (Liu et al., 2015). Automated systems also make it easier to implement sustainable practices such as real-time monitoring of energy usage and waste generation, enabling manufacturers to identify inefficiencies and make adjustments in real-time.

The adoption of robotic systems in manufacturing also plays a significant role in reducing emissions. These robots are designed to work efficiently and often use less energy than traditional manual labor, contributing to lower overall energy consumption. Moreover, automation allows for the use of more sustainable materials and manufacturing methods, further enhancing the environmental benefits of green manufacturing. As industries continue to embrace automation, it will play an increasingly important role in driving the shift towards more sustainable and resource-efficient production systems (Vermesan et al., 2014).

Advancing Solar Power: The Future of Clean Energy

Solar power is one of the most promising renewable energy sources due to its abundance and versatility. Advances in solar technology have made it more efficient and cost-effective than ever before, paving the way for widespread adoption. Photovoltaic (PV) panels, which convert sunlight into electricity, have seen significant improvements in energy conversion rates and durability. In addition to traditional rooftop solar panels, newer technologies such as concentrated solar power (CSP) and thin-film solar cells offer innovative ways to harness solar energy in diverse applications (Jacobson et al., 2015).

Solar power not only provides a clean alternative to fossil fuels but also offers significant economic benefits. By reducing dependency on imported energy, solar power can enhance energy security and create jobs in the renewable energy sector. Additionally, large-scale solar farms and residential solar installations can help reduce energy costs, benefiting both consumers and businesses. As solar power continues to evolve and become more accessible, it will play a crucial role in achieving global climate goals and transitioning to a low-carbon economy (NREL, 2020).

Green Technology and Global Climate Action

Green technology plays a central role in global climate action by providing the tools and solutions necessary to mitigate climate change. From renewable energy systems to energy-efficient appliances, green technologies can help reduce greenhouse gas emissions and limit the effects of global warming. The transition to clean energy sources such as wind, solar, and hydroelectric power is one of the most effective ways to decarbonize the energy sector. Additionally, innovations in energy storage, carbon capture, and sustainable agriculture are essential in addressing climate change and ensuring food and water security in a warming world.

Governments, businesses, and individuals must continue to invest in and promote green technologies to achieve climate goals and build a sustainable future. Public policies that incentivize the adoption of green technologies, such as tax credits for solar panels and electric vehicles, will help accelerate the transition to a low-carbon economy. As the world faces an increasingly urgent climate crisis, green technology remains a powerful tool for combating environmental degradation and ensuring a livable planet for future generations (UNFCCC, 2020).

Chapter 13: Legislation and Policy: The role of governments

Introduction to Government Legislation and Policy

Governments play a crucial role in shaping societal norms and addressing pressing global issues through legislation and policy. Legislation refers to the laws enacted by governmental bodies, while policy encompasses the strategies and approaches designed to manage public issues. Both are essential tools for achieving social, environmental, and economic goals, from promoting public health to addressing climate change. By creating regulatory frameworks, governments ensure the smooth functioning of societies and foster sustainable development (Baldwin, Cave, & Lodge, 2012).

The role of governments in enacting effective legislation and policy is vital for the success of any societal issue. In the context of environmental protection, for example, governments must balance the needs of economic development with environmental sustainability. Policies such as the Paris Agreement reflect international efforts to tackle climate change through coordinated action (UNFCCC, 2015).

Governments must also ensure that policies reflect the current realities and emerging challenges in society. In order to do so, they must actively engage with experts, stakeholders, and the public to craft inclusive and effective legislation. With climate change and other global crises gaining prominence, the responsibility of governments to implement actionable and impactful policy is more crucial than ever (Howlett, Ramesh, & Perl, 2016).

The Importance of Government Regulation

Regulation is a cornerstone of government action. It ensures that societal behaviors align with public interests, particularly when individual actions can negatively impact the community. Effective regulation can address issues such as environmental protection, public health, and safety by establishing clear legal standards. Governments regulate industries to mitigate harmful practices, promote fairness, and ensure long-term sustainability (Jordana & Levi-Faur, 2005).

In environmental regulation, for example, governments set limits on pollutants, enforce waste management practices, and implement standards for sustainable resource use. Laws like the Clean Air Act and Clean Water Act in the United States have significantly reduced air and water pollution over the years, demonstrating the power of government regulation in protecting natural resources (US EPA, 2021). By adopting stringent regulations, governments promote the responsible use of resources and incentivize industries to innovate in greener technologies.

The role of government regulation also extends to protecting public health and safety. For instance, policies governing food safety, workplace conditions, and public health services prevent the spread of diseases, improve quality of life, and reduce social costs. In this way, government regulation is not just a tool for promoting public safety but also a means of fostering a fair and just society (Baldwin et al., 2012).

Policy Development and Government Action

Policy development is an essential function of government, as it determines how resources are allocated and how societal challenges are addressed. Policymaking involves a complex process of identifying problems, designing solutions, and implementing strategies that reflect the needs of citizens. Governments rely on empirical evidence, expert advice, and public input to craft policies that are both practical and effective. Policy development also requires balancing competing interests, such as economic growth, social equity, and environmental protection (Howlett et al., 2016).

The process of policy development is iterative and often involves multiple stages. Initially, governments identify pressing issues that require intervention, such as climate change, public health crises, or economic instability. Then, policymakers propose solutions and evaluate the potential impact of various strategies. Public consultations and stakeholder engagement play a critical role in ensuring that policies meet the needs of diverse groups. Finally, after implementation, policies are continually assessed and adjusted based on feedback and outcomes (Birkland, 2016).

Governments' ability to develop and enact effective policies is fundamental to addressing the challenges of the modern world. With issues such as climate change and global inequality, governments must act swiftly and decisively to protect citizens and future generations.

Environmental Legislation and Sustainable Development

Environmental legislation plays a pivotal role in shaping sustainable development policies. It establishes the legal framework for protecting natural resources and mitigating environmental degradation. Governments use environmental laws to regulate industrial emissions, waste disposal, deforestation, and biodiversity conservation. Through effective environmental legislation, governments can strike a balance between economic development and environmental protection.

For example, the establishment of protected areas and national parks through environmental laws ensures the conservation of biodiversity and the preservation of ecosystems. Additionally, international agreements such as the Kyoto Protocol and the Paris Agreement emphasize the importance of global collaboration in addressing climate change through binding commitments from nations to reduce greenhouse gas emissions (UNFCCC, 2015).

Environmental legislation also fosters innovation by encouraging businesses to develop green technologies and sustainable practices. The imposition of carbon taxes or cap-and-trade systems incentivizes industries to adopt cleaner technologies, thus reducing their environmental footprint. Governments that enforce strict environmental regulations not only safeguard natural resources but also promote long-term economic growth through sustainable practices (Tietenberg & Lewis, 2016).

International Policies and Global Cooperation

In an increasingly interconnected world, international policies and agreements are essential for addressing global challenges that transcend national borders. Issues such as climate change, biodiversity loss, and transboundary pollution require coordinated global action. Through international treaties and conventions, governments come together to establish common standards, share knowledge, and collaborate on solutions to global environmental issues (Keohane & Victor, 2011).

For example, the United Nations Framework Convention on Climate Change (UNFCCC) and its associated agreements, such as the Kyoto Protocol and the Paris Agreement, provide a framework for countries to negotiate emissions reductions and other climate-related goals. These agreements foster cooperation among nations and ensure that all countries, regardless of their economic status, contribute to global efforts to mitigate climate change (UNFCCC, 2015).

International policies also facilitate the exchange of technological innovations and financial resources between developed and developing nations, helping the latter to transition to cleaner, more sustainable energy systems. In this way, international cooperation is essential for achieving global sustainability and ensuring the equitable distribution of resources (Falkner, 2016).

Economic Policies and Sustainable Growth

Economic policies are integral to fostering sustainable growth, as they guide how nations manage their resources, investments, and industries. Governments use economic policies to promote social welfare, reduce inequality, and stimulate economic activity while ensuring environmental sustainability. A key challenge in economic policy is balancing the pursuit of growth with the need to protect the environment and address social inequalities (Dasgupta, 2021).

For instance, governments can incentivize sustainable economic activities through subsidies for green technologies, tax credits for renewable energy projects, and investment in clean infrastructure. Policies such as carbon pricing and the implementation of green taxes can also discourage harmful practices and promote the transition to low-carbon economies. By integrating sustainability into economic decision-making, governments can foster long-term economic prosperity without compromising the health of the planet (Stern, 2007).

The concept of "green growth" emphasizes the possibility of achieving economic growth while ensuring the responsible use of natural resources. Countries such as Denmark and Germany have successfully integrated green growth strategies into their economic policies, leading to the expansion of renewable energy sectors and the creation of green jobs (OECD, 2017). These policies demonstrate the potential of economic measures to support sustainable development.

The Role of Public Participation in Policy Formation

Public participation is an essential element of effective policy formation. Governments that engage citizens in decision-making processes ensure that policies reflect the needs and preferences of the people they serve. Public participation not only promotes transparency but also enhances the legitimacy of policies and fosters democratic values (Arnstein, 1969).

In environmental policy, for example, public consultations provide an opportunity for stakeholders to voice their concerns about proposed regulations, such as land-use plans or climate action initiatives. These consultations enable policymakers to incorporate local knowledge and diverse perspectives into their decision-making process, leading to more equitable and effective policies (Gunningham et al., 2004).

Participatory approaches also contribute to social cohesion, as citizens feel more invested in policies that they helped shape. By involving communities in environmental governance, governments can build public support for sustainable practices, increase awareness about environmental issues, and promote collective action (Dovers & Hezri, 2010).

Social Justice and Government Policies

Social justice and equity are central to many government policies, particularly in addressing environmental issues. Environmental justice seeks to ensure that all people, regardless of race, ethnicity, or socioeconomic status, have access to clean air, water, and land. Governments play a critical role in addressing inequalities by enacting policies that protect vulnerable populations from environmental hazards (Bullard, 2005).

For example, in response to growing concerns about pollution in low-income communities, governments may implement policies to reduce exposure to toxic substances, enforce stricter regulations on industrial emissions, and improve public health infrastructure. Furthermore, policies aimed at addressing environmental racism—such as disproportionately high levels of pollution in marginalized communities—are essential for promoting fairness and inclusion (Mohai, Pellow, & Roberts, 2009).

By prioritizing social justice in environmental policies, governments can help ensure that all citizens benefit from sustainable development and have a say in the decisions that affect their lives. This approach strengthens social cohesion and creates a more equitable society (Walker, 2012).

International Trade and Environmental Policy

International trade has significant implications for environmental sustainability. As global trade expands, so too do the environmental impacts associated with the transportation of goods and the exploitation of natural resources. Governments must strike a balance between encouraging trade to foster economic growth and implementing environmental policies that protect global ecosystems (Copeland & Taylor, 2004).

Trade policies that encourage the adoption of environmental standards, such as eco-labeling, carbon footprint tracking, and product lifecycle assessments, can promote sustainable production and consumption. By incorporating environmental considerations into trade agreements, such as the World Trade Organization (WTO) and regional trade pacts, governments can ensure that trade practices align with sustainability goals (Ostry et al., 2009).

Additionally, international trade policies must account for the environmental impacts of supply chains, including deforestation, overfishing, and pollution. Governments can enact regulations to ensure that industries adopt sustainable practices and that trade does not come at the expense of the environment. By doing so, governments can promote a green economy that balances economic growth with environmental protection (Green, 2017).

Climate Change Policy and Government Responsibility

Governments worldwide have a critical role to play in addressing climate change through the development of comprehensive climate policies. Climate change represents one of the most significant threats to global ecosystems, human health, and economies. As such, it is imperative that governments take bold action to mitigate greenhouse gas emissions and adapt to the changing climate.

International agreements such as the Paris Agreement highlight the global commitment to limiting global temperature rise to below 2°C above pre-industrial levels. Governments that ratify such agreements are responsible for setting and achieving national targets for emissions reductions. These targets often require significant policy reforms across sectors such as energy, transportation, and agriculture (UNFCCC, 2015).

At the national level, governments can implement climate policies that promote renewable energy, energy efficiency, carbon pricing, and sustainable land use. Policies like carbon taxes or cap-and-trade systems provide economic incentives for industries to reduce emissions. Additionally, governments must ensure that vulnerable communities have the resources and infrastructure to adapt to climate impacts, such as sea-level rise and extreme weather events (Stern, 2007).

The Role of Governments in Environmental Justice, Climate Action, Green Innovation, and Beyond

Continuing from this point, each subsequent page will delve deeper into specific areas such as environmental justice, the promotion of green innovation, economic policies for sustainable development, global cooperation on environmental issues, and the importance of robust government action across different sectors. I'll ensure the structure remains aligned with your request, with appropriate in-citations and references.

The Role of Legislation in Promoting Sustainable Development

Legislation serves as the foundation for achieving sustainable development by creating legal frameworks that guide and regulate human activities. Governments implement laws that ensure environmental protection, equitable distribution of resources, and social welfare. In the context of sustainable development, legislation must balance economic growth, social equity, and environmental conservation (Meadows et al., 2004).

Sustainable development legislation includes laws that regulate resource extraction, energy production, waste management, and environmental conservation. For example, the Environmental Protection Agency (EPA) in the United States enforces regulations designed to protect air and water quality, limit industrial emissions, and encourage the development of cleaner technologies (US EPA, 2021). Similarly, the European Union's Circular Economy Action Plan promotes waste reduction, recycling, and resource efficiency, supporting the transition to a more sustainable economic model (European Commission, 2020).

At the international level, sustainable development goals (SDGs) adopted by the United Nations (UN) provide a global framework for governments to align their policies with sustainability objectives. These goals, which include eradicating poverty, achieving gender equality, and addressing climate change, guide national legislation in various countries toward fostering a sustainable future (United Nations, 2015).

Through effective legislation, governments can not only manage the impact of human activities on the environment but also promote social equity and economic stability. Sustainable development requires a comprehensive approach that integrates environmental, social, and economic considerations into the legal and policy frameworks that guide governmental decision-making.

Policies for Mitigating Environmental Impacts

Governments play a crucial role in mitigating the environmental impacts of industrialization and urbanization. Environmental policies aimed at reducing the harmful effects of human activities on ecosystems and natural resources are central to achieving sustainability goals. Key policy areas include carbon emissions reduction, air and water quality standards, and land use planning (Perman et al., 2011).

Carbon pricing mechanisms, such as carbon taxes and cap-and-trade systems, incentivize industries to reduce greenhouse gas emissions by placing a financial cost on carbon emissions. For example, the European Union Emissions Trading System (EU ETS) allows companies to trade emission allowances, encouraging them to reduce their carbon footprint to remain economically competitive (Tietenberg & Lewis, 2016). These policies provide market-driven solutions to combat climate change while fostering innovation in clean technologies.

Governments also use direct regulatory measures to limit environmental pollution. The Clean Water Act and Clean Air Act in the United States are examples of legislation that set standards for air and water quality. These regulations help prevent industrial discharge of pollutants into the environment, reducing the risk of contamination and promoting public health (US EPA, 2021).

Finally, land-use policies such as zoning laws and protected area designations can mitigate environmental degradation by controlling urban sprawl and conserving biodiversity. Sustainable land-use practices ensure that ecosystems are preserved for future generations while supporting the needs of growing populations.

Governance and Accountability in Environmental Protection

Governance and accountability are central to ensuring that environmental policies are implemented effectively. Governments must not only design policies but also monitor and enforce compliance with environmental laws and regulations. Transparency, public participation, and accountability mechanisms are essential for fostering trust and ensuring that policies achieve their intended outcomes (Andonova, 2017).

One of the main challenges in environmental governance is ensuring that industries, businesses, and local governments adhere to environmental laws. Effective monitoring systems and enforcement mechanisms are necessary to detect violations and impose penalties for non-compliance. Governments can also establish environmental agencies tasked with overseeing the implementation of policies and ensuring that regulations are followed (Meadowcroft, 2007).

Public participation in environmental governance also plays an important role in holding governments accountable. Citizens can engage in public consultations, participate in local environmental planning, and hold policymakers accountable for the environmental outcomes of their decisions. Additionally, non-governmental organizations (NGOs) and civil society groups often serve as watchdogs, advocating for stronger environmental protection laws and greater transparency in government decision-making (Betsill & Corell, 2008).

In a globalized world, international governance frameworks also provide mechanisms for holding governments accountable for their environmental commitments. For instance, the Paris Agreement includes a transparency framework that allows countries to report on their progress in reducing greenhouse gas emissions, providing accountability at the international level (UNFCCC, 2015).

The Role of Governments in Promoting Green Innovation

Green innovation refers to the development of new technologies, processes, and practices that contribute to environmental sustainability. Governments have a critical role to play in fostering green innovation through policies that incentivize research and development (R&D), support green entrepreneurship, and facilitate the transition to low-carbon economies.

One way governments promote green innovation is by providing financial support for R&D in renewable energy, waste management, and clean technologies. By funding research projects and offering grants to innovators, governments help accelerate the development of new technologies that can address environmental challenges (Nelson & Winter, 1982).

Furthermore, governments can create market incentives for green innovation by implementing policies that encourage sustainable practices. For instance, subsidies for renewable energy production, tax credits for electric vehicles, and carbon pricing mechanisms make green technologies more economically viable and attractive to businesses and consumers (Grubb et al., 2014).

Governments also play a key role in creating favorable conditions for green entrepreneurship. By establishing green procurement policies, governments can stimulate demand for eco-friendly products and services. In this way, governments can create markets for green innovation and foster a culture of sustainability.

The Importance of International Cooperation for Environmental Policy

Environmental challenges, such as climate change, biodiversity loss, and deforestation, often transcend national borders. As a result, international cooperation is critical to addressing global environmental issues. Governments must work together to create and implement policies that promote sustainability and mitigate the impacts of environmental degradation (Keohane & Victor, 2011).

International environmental agreements, such as the Paris Agreement on climate change and the Convention on Biological Diversity, exemplify the importance of collective action in addressing global challenges. Through these agreements, countries commit to common goals and share knowledge and resources to achieve environmental objectives. The Paris Agreement, for example, aims to limit global temperature rise and enhance countries' abilities to adapt to climate change (UNFCCC, 2015).

International cooperation also enables countries to share best practices and technologies for environmental management. For instance, developed countries can support developing nations in transitioning to renewable energy through financial assistance and technological transfer. In this way, international cooperation fosters equity and helps to bridge the gap between developed and developing nations in the global pursuit of environmental sustainability (Falkner, 2016).

Government Leadership in Environmental Protection

Governments are often the driving forces behind environmental protection efforts, providing the leadership needed to tackle complex environmental issues. By enacting policies, regulations, and laws that prioritize environmental sustainability, governments set the direction for industries, communities, and individuals to follow (Gunningham et al., 2004).

In addition to legislation, governments can lead by example. By adopting sustainable practices within their own operations—such as reducing energy consumption, adopting renewable energy sources, and promoting sustainable procurement—governments demonstrate their commitment to environmental stewardship. Public sector leadership can inspire others to adopt similar practices and contribute to the broader goal of sustainability.

Governments also have a responsibility to lead in international fora, negotiating agreements and setting international standards that promote global environmental protection. Through diplomatic channels and international organizations such as the United Nations, governments can advocate for stronger environmental protections and greater international cooperation (Keohane & Victor,

Government Accountability in Environmental Policy Enforcement

Accountability in the enforcement of environmental policies is crucial for ensuring that governments meet their environmental goals and commitments. While governments may pass legislation and develop policies aimed at sustainability, without strong enforcement mechanisms, these efforts can fall short. Effective monitoring, transparent reporting, and the imposition of penalties for non-compliance are essential components of a robust environmental governance framework (Meadowcroft, 2007).

Governments must ensure that laws regulating air and water quality, waste management, and resource extraction are not only enacted but also monitored for compliance. Independent regulatory agencies, such as the Environmental Protection Agency (EPA) in the United States, are tasked with monitoring industry practices, conducting inspections, and imposing fines on companies that violate environmental regulations (US EPA, 2021). These agencies also often provide public reports on environmental conditions, offering transparency and ensuring that the public is aware of governmental efforts and outcomes.

In addition to traditional regulatory enforcement, some governments are increasingly adopting market-based instruments such as carbon pricing, tradable permits, and subsidies to incentivize environmental compliance (Tietenberg & Lewis, 2016). These economic tools complement regulatory measures by creating financial incentives for businesses and individuals to reduce their environmental impact.

The Role of Local Governments in Environmental Protection

While national governments are often responsible for the overarching environmental policies and regulations, local governments play a critical role in implementing sustainable practices and addressing environmental issues at the grassroots level. Local governments are more closely connected to the communities they serve, allowing them to address local environmental challenges in ways that reflect the unique needs of their populations (Barrett & Graddy, 2000).

Cities and municipalities have the power to create and enforce local environmental laws, zoning regulations, and building codes. For example, many cities have adopted local air quality standards and sustainable transportation policies that reduce traffic congestion and promote public transit or electric vehicle usage. Local governments can also encourage the use of renewable energy sources through zoning laws and incentives for solar and wind energy installations (Koch, 2015).

Furthermore, local governments often play a crucial role in the management of green spaces, urban forests, and local ecosystems. Many municipalities have established urban gardens, parks, and protected areas that contribute to biodiversity conservation and environmental education. Local environmental policies can also address waste management, water conservation, and climate resilience, ensuring that communities are better prepared to cope with environmental challenges.

Conclusion: The Importance of Government Action in Environmental Protection

In conclusion, the role of governments in environmental protection and sustainability is indispensable. From creating and enforcing environmental laws to fostering green innovation and international cooperation, governments are at the forefront of addressing global environmental challenges. Through effective legislation, transparent governance, and accountability, governments can ensure that environmental policies are not only ambitious but also actionable and impactful.

As the global community faces urgent environmental issues such as climate change, biodiversity loss, and resource depletion, government action becomes even more critical. Governments must lead by example, engage in international dialogue, and empower local communities to build a sustainable future for all. By working collaboratively with businesses, civil society, and international organizations, governments can achieve significant progress toward the environmental goals set forth by the global community.

Environmental protection is not a singular effort but a shared responsibility. Governments have the power to catalyze systemic change by implementing the policies, regulations, and frameworks that drive sustainability. As the global population continues to grow and face new challenges, governments must remain committed to a future that prioritizes the health of the planet, the well-being of its inhabitants, and the preservation of natural resources for generations to come.

Chapter 14: Corporate Responsibility: Business and the environment

Introduction to Corporate Responsibility and the Environment

Corporate responsibility refers to the ethical obligations that businesses have towards society and the environment. Businesses are increasingly expected to go beyond profit generation and consider their environmental and social impact. As the world faces challenges such as climate change, biodiversity loss, and resource depletion, corporations play a pivotal role in addressing these issues. In the context of environmental sustainability, corporate responsibility involves adopting practices that reduce environmental harm and promote long-term ecological balance (Carroll, 1999).

In recent years, there has been growing recognition that corporate actions can significantly impact environmental outcomes. Environmental degradation resulting from industrial activities, such as pollution, deforestation, and greenhouse gas emissions, has led to calls for more sustainable business practices (Hart, 1995). As a result, businesses are increasingly being held accountable not only by regulatory bodies but also by consumers, investors, and employees who demand greater environmental stewardship.

Corporate responsibility encompasses a wide range of activities, including reducing carbon footprints, conserving natural resources, promoting renewable energy, and ensuring ethical supply chains. These practices are part of a broader trend towards "corporate social responsibility" (CSR), which emphasizes a company's role in supporting social and environmental causes (Porter & Kramer, 2006).

In this context, businesses can become active participants in promoting sustainability by integrating environmental considerations into their operations and decision-making processes. This shift towards more responsible business practices is not only beneficial for the environment but also offers long-term economic benefits, including enhanced brand reputation and consumer loyalty (Kotler & Lee, 2005).

The Business Case for Corporate Environmental Responsibility

Integrating environmental responsibility into business operations offers tangible benefits, making it an essential aspect of modern corporate strategy. Companies that embrace sustainable practices can reduce costs, mitigate risks, and enhance their brand image, ultimately contributing to long-term profitability (McKinsey & Company, 2011). In addition, environmental responsibility can foster innovation, enabling businesses to develop new products and services that meet growing consumer demand for eco-friendly solutions.

Cost savings from sustainable practices are particularly evident in energy efficiency initiatives. For example, by reducing energy consumption, businesses can lower operating costs while simultaneously reducing their carbon footprints. According to the Carbon Trust (2014), companies that implement energy efficiency measures can achieve significant cost savings, especially in sectors with high energy usage such as manufacturing and transportation.

Moreover, businesses that prioritize environmental responsibility are better equipped to navigate the risks associated with climate change and resource scarcity. For instance, companies that rely on natural resources may face disruptions due to supply chain challenges, increasing operational costs and impacting production (Sullivan & Gouldson, 2017). By adopting sustainable practices, businesses can ensure that their operations are more resilient to environmental and regulatory changes.

A proactive environmental strategy can also enhance a company's reputation, attracting environmentally conscious consumers and investors. A strong commitment to sustainability can lead to greater brand loyalty and open new markets, particularly as consumers become more aware of the environmental impact of their purchasing decisions (Nielsen, 2015).

Environmental Regulation and Corporate Responsibility

Environmental regulations are one of the primary mechanisms through which governments seek to manage the impact of businesses on the environment. Regulations set standards for pollution control, waste management, and resource conservation, thereby holding companies accountable for their environmental impact. However, these regulations are often seen by businesses as constraints. In recent years, businesses have recognized that proactive compliance with environmental laws can lead to competitive advantages (Porter & Van der Linde, 1995).

The notion of "environmental regulation as innovation" suggests that regulations can stimulate innovation by challenging businesses to develop new technologies and processes that reduce environmental harm (Porter & Van der Linde, 1995). For example, strict emissions standards have driven the automotive industry to invest heavily in electric vehicle technology, resulting in a growing market for low-emission vehicles. Similarly, regulations targeting waste reduction have encouraged businesses to develop more efficient recycling and resource recovery systems.

While regulatory compliance is crucial, businesses that go beyond the minimum requirements can gain additional benefits. Environmental leadership positions companies as responsible actors in the eyes of consumers and investors, often leading to stronger financial performance. Companies that engage in voluntary environmental initiatives or achieve certifications such as ISO 14001 demonstrate their commitment to sustainability and differentiate themselves in the marketplace (Hart, 1995).

Corporate Social Responsibility and the Environment

Corporate social responsibility (CSR) is an essential part of the corporate strategy that focuses on creating positive social and environmental impacts. CSR has evolved over time to encompass more than just philanthropy, evolving into a critical business practice that integrates environmental and social considerations into business decision-making (Elkington, 1997). Companies with strong CSR programs engage in activities that benefit society, including supporting environmental conservation initiatives, reducing carbon emissions, and promoting sustainable resource management.

One of the key environmental aspects of CSR is the commitment to reducing environmental impacts through sustainable practices. Many companies now aim to reduce their carbon footprint, limit waste generation, and conserve resources. For instance, companies like Unilever and Nestlé have committed to reducing their environmental impact by adopting more sustainable sourcing practices and packaging innovations (Unilever, 2020). These efforts are not just about compliance with regulations but are part of broader corporate values that align with global environmental goals such as the United Nations' Sustainable Development Goals (SDGs).

Moreover, CSR programs often involve corporate transparency, where businesses disclose their environmental impact through sustainability reports. This openness builds trust with stakeholders and encourages companies to continuously improve their environmental performance (Kolk & van Tulder, 2002). Companies that demonstrate transparency and take meaningful action on environmental issues gain a competitive advantage in today's increasingly sustainability-conscious market.

Greenwashing and Its Impact on Corporate Responsibility

While many companies are making significant strides toward sustainability, there are concerns about the phenomenon of "greenwashing." Greenwashing refers to the practice of companies marketing themselves as environmentally friendly without actually making substantial efforts to reduce their environmental impact (Lyon & Montgomery, 2015). This can mislead consumers and investors who are trying to support sustainable businesses. As companies increasingly claim to be "green," it becomes essential to differentiate between those that genuinely prioritize sustainability and those that use environmental claims as a marketing strategy.

Greenwashing undermines trust in corporate responsibility efforts and can hinder progress towards environmental goals. For instance, companies that make unsubstantiated claims about their products being "eco-friendly" may attract customers based on misleading information, creating a false sense of progress in sustainability. This not only damages the company's reputation when the truth is revealed but can also lead to regulatory scrutiny and legal consequences (Delmas & Burbano, 2011).

To combat greenwashing, businesses must commit to transparency and provide clear, verifiable evidence of their sustainability efforts. Third-party certifications, such as the Fair Trade or Energy Star labels, can help consumers identify genuinely sustainable products. Furthermore, regulatory frameworks and watchdog organizations are increasingly holding businesses accountable for greenwashing, ensuring that companies meet their sustainability claims with concrete actions.

Corporate Sustainability Reporting

Sustainability reporting has become an integral part of corporate responsibility, as it allows businesses to disclose their environmental, social, and governance (ESG) performance to stakeholders. By publishing regular sustainability reports, companies provide transparency regarding their impact on the environment and demonstrate their commitment to sustainability (Kolk, 2008). These reports typically include information on a company's energy consumption, waste management practices, carbon emissions, and resource usage.

The Global Reporting Initiative (GRI) is one of the most widely adopted frameworks for sustainability reporting, providing companies with standardized guidelines to measure and disclose their sustainability efforts. GRI's guidelines help businesses communicate their environmental goals and progress clearly and credibly. In addition to GRI, there are other reporting frameworks like the Carbon Disclosure Project (CDP) and the Sustainability Accounting Standards Board (SASB) that offer standardized methods for measuring environmental performance (Eccles & Krzus, 2018).

Transparency in sustainability reporting is essential for building trust with stakeholders, including investors, customers, and employees. It also enables businesses to identify areas for improvement and take corrective actions to reduce their environmental footprint. However, it is important for companies to ensure that their sustainability reports are accurate, comprehensive, and not overstated. This can be achieved by independent audits and third-party verification to ensure the integrity of the information shared.

Circular Economy and Corporate Responsibility

A circular economy is an innovative economic model that aims to reduce waste and promote the continuous use of resources. Unlike the traditional linear model of "take, make, dispose," the circular economy focuses on reusing, recycling, and regenerating materials to extend the lifecycle of products and minimize environmental impact (Ellen MacArthur Foundation, 2013). Many companies are adopting circular economy principles to reduce waste, decrease resource consumption, and improve sustainability. This shift aligns with growing consumer expectations for companies to take responsibility for the environmental impact of their products throughout their entire lifecycle.

Incorporating circular economy practices into business operations can have significant environmental and economic benefits. For instance, by using renewable resources, businesses can reduce their reliance on finite natural resources and decrease the energy needed to manufacture new products (Bocken et al., 2016). Companies like Patagonia and IKEA have already embraced circular economy models by focusing on product repairability, recycling, and promoting the reuse of materials. This transition not only reduces waste but also helps businesses build brand loyalty among environmentally conscious consumers.

The circular economy also presents opportunities for innovation. Businesses that embrace circular principles can develop new business models that create value from waste. For example, companies can design products for disassembly, allowing for easier recycling and reducing the need for raw materials. By adopting circular practices, companies can contribute to the broader goal of achieving a sustainable economy that benefits both the environment and society.

Corporate Responsibility and Climate Change Mitigation

As global concern about climate change intensifies, businesses are being called upon to play a critical role in reducing greenhouse gas emissions. Corporate responsibility in the context of climate change mitigation refers to the actions taken by companies to reduce their carbon footprint and support global efforts to address climate change. This includes reducing energy consumption, improving energy efficiency, and investing in low-carbon technologies (Hart, 1995).

In recent years, many companies have committed to setting science-based targets for carbon reduction, which are aligned with the goals of the Paris Agreement. These targets help businesses to set measurable and ambitious goals for reducing emissions. For example, companies like Google, Apple, and Microsoft have made significant strides toward achieving carbon neutrality through the use of renewable energy sources, energy-efficient technologies, and carbon offset programs (CDP, 2020).

Moreover, businesses that focus on climate change mitigation also recognize the financial risks posed by climate change. Extreme weather events, resource scarcity, and regulatory changes can all impact a company's operations, supply chain, and market conditions. By proactively addressing these risks, companies can better prepare for the future while contributing to global efforts to reduce the severity of climate change (Kolk & Pinkse, 2005).

Companies can also play a significant role in influencing government policy and shaping climate change legislation. By advocating for stronger climate regulations and participating in climate initiatives, businesses can contribute to creating a more sustainable global economy.

Corporate Responsibility in the Supply Chain

The supply chain plays a significant role in a company's environmental impact. A sustainable supply chain integrates environmental considerations into every stage, from procurement to production, distribution, and disposal. Corporations that are serious about environmental responsibility must ensure that their suppliers adhere to sustainable practices as well. This can include reducing emissions, using sustainable materials, and promoting ethical labor practices (Seuring & Müller, 2008).

An important aspect of sustainable supply chain management is traceability. Many companies are increasingly transparent about the origins of their materials and the environmental impact of their supply chains. For instance, companies in the fashion and electronics industries have worked with suppliers to ensure that raw materials are sourced responsibly and that production processes minimize waste and pollution (Fletcher, 2008). By establishing clear expectations and collaborating with suppliers to meet sustainability targets, businesses can reduce the environmental footprint of their supply chains.

Corporate responsibility also extends to how businesses manage their waste. Companies that manage their supply chains sustainably ensure that waste is minimized, reused, or recycled whenever possible. By working with third-party auditors and certifications, companies can verify that their supply chain partners are following the same high environmental standards. This collaboration helps businesses maintain a competitive edge by appealing to consumers who value sustainability and ethical sourcing.

Environmental Education in Corporate Strategy

Environmental education within corporate strategy is essential for fostering a culture of sustainability across all levels of a business. As environmental issues become increasingly complex, it is crucial for employees, leaders, and stakeholders to have the knowledge and tools to make informed decisions about sustainability. Many corporations now provide training programs on environmental issues, such as climate change, waste management, and renewable energy (Senge et al., 2008).

Corporate environmental education not only equips employees with the skills to implement sustainable practices but also aligns the company's values with broader social and environmental goals. For example, companies like Tesla and Google have made environmental education a core part of their business models, integrating sustainability into their internal training programs and leadership development (Wheeler & Sillanpää, 1997).

Moreover, environmental education helps to create a sense of ownership among employees, who are encouraged to contribute to sustainability initiatives both within and outside the company. By integrating environmental education into corporate strategy, businesses can foster innovation and collaboration, leading to more effective and efficient sustainability practices.

Environmental education also plays a role in influencing consumer behavior. Companies that educate consumers on environmental issues, such as the benefits of recycling or using energy-efficient products, empower them to make better choices. This not only improves the environmental performance of society but also strengthens the company's position as a responsible corporate citizen.

Eco-Innovation and Corporate Responsibility

Eco-innovation refers to the development of new products, processes, or business models that contribute to environmental sustainability while providing value to consumers and businesses. This concept is central to corporate responsibility, as companies that innovate in environmentally friendly ways not only reduce their ecological impact but also gain a competitive advantage in the marketplace. Eco-innovative companies are more likely to meet regulatory requirements, attract environmentally conscious consumers, and achieve long-term profitability (Porter & van der Linde, 1995).

The process of eco-innovation can involve the creation of new technologies, the development of environmentally friendly products, or the implementation of resource-efficient production processes. For instance, companies like Tesla have revolutionized the electric vehicle industry by creating eco-friendly alternatives to traditional gasoline-powered cars, while companies like Interface have pioneered sustainable carpet tiles that use recycled materials and reduce waste (Nidumolu et al., 2009).

Eco-innovation is not only limited to product development but also includes operational improvements. Companies are increasingly focusing on enhancing energy efficiency, reducing waste, and minimizing water usage in their manufacturing processes. By integrating eco-innovation into business strategies, companies can reduce costs and achieve environmental goals. Additionally, eco-innovations can open up new markets and create additional value for both businesses and society.

To support eco-innovation, governments and regulatory bodies play an essential role by offering incentives, grants, and policies that encourage sustainable business practices. As eco-innovation becomes a key driver of corporate responsibility, businesses will continue to shape the future of sustainability.

Green Supply Chains and Corporate Responsibility

Green supply chains are an important aspect of corporate responsibility, focusing on minimizing the environmental impact of goods and services throughout their lifecycle. This includes everything from sourcing raw materials to product manufacturing, distribution, and disposal. Green supply chain management seeks to integrate sustainable practices into each stage of the supply chain, reducing waste, energy use, and emissions (Srivastava, 2007).

To implement a green supply chain, companies need to work closely with suppliers to ensure that sustainable practices are followed at every level. This may involve choosing suppliers that use renewable resources, ensuring that transportation methods are energy-efficient, and reducing packaging waste. Companies like Walmart have been successful in implementing green supply chains by requiring suppliers to meet sustainability standards and sharing best practices across their network (Busse et al., 2017).

One of the main benefits of a green supply chain is the reduction of costs through increased efficiency. By reducing waste, energy consumption, and materials usage, businesses can lower production costs and improve their environmental performance. Furthermore, adopting green supply chain practices can improve a company's reputation, attracting customers and investors who prioritize sustainability.

Green supply chains also promote innovation. As companies search for new ways to reduce environmental impacts, they often discover innovative solutions that improve not only sustainability but also operational efficiency. By embracing green supply chains, businesses can reduce their environmental footprint while maintaining profitability.

Corporate Responsibility in the Food Industry

The food industry has a significant environmental impact, encompassing issues such as greenhouse gas emissions, water usage, land degradation, and waste. In response, companies in the food sector have increasingly embraced corporate responsibility by adopting sustainable practices that reduce their environmental footprint. Corporate responsibility in the food industry focuses on the entire lifecycle of food products, from sourcing ingredients to packaging, distribution, and consumption (Garnett, 2013).

One key aspect of corporate responsibility in the food industry is sustainable sourcing. Companies are increasingly sourcing ingredients that are produced in environmentally friendly ways, such as organic farming, fair trade certification, and responsible fishing practices. For example, Starbucks has committed to sourcing 100% of its coffee beans from ethical and sustainable sources, while Unilever has pledged to ensure that all of its agricultural raw materials are sustainably sourced by 2020 (Unilever, 2020).

Reducing food waste is another critical focus for companies in the food industry. The Food and Agriculture Organization (FAO) estimates that approximately one-third of food produced globally is wasted. Companies like Tesco and Carrefour have implemented strategies to reduce food waste through better inventory management, partnerships with food banks, and promoting sustainable consumption practices (Parfitt et al., 2010).

Additionally, many food companies are working to improve the environmental sustainability of their packaging. Companies like Nestlé and Coca-Cola have made commitments to reduce plastic packaging and increase the use of recyclable materials. By addressing both environmental and social challenges, the food industry is taking significant steps toward corporate responsibility.

Collaboration Between Government and Corporations for Sustainability

Achieving environmental sustainability requires collaboration between governments, corporations, and other stakeholders. While businesses can drive change by adopting sustainable practices, governments play a critical role in creating an enabling environment that promotes sustainability. Government policies, regulations, and incentives can guide corporate behavior toward sustainability by setting standards for emissions, waste management, and resource conservation (Porter & van der Linde, 1995).

For example, many countries have enacted policies that encourage businesses to invest in renewable energy. Governments may offer tax incentives or grants for businesses that adopt renewable energy technologies, such as solar panels or wind turbines. These policies not only reduce the financial barriers to implementing sustainable practices but also signal to the market that sustainability is a priority (Gunningham et al., 2004).

Additionally, governments can work with businesses to create long-term sustainability strategies. By forming partnerships with the private sector, governments can promote innovative solutions and share the costs of implementing environmental technologies.

Collaborative efforts can also lead to the creation of environmental standards and certifications that help businesses meet sustainability goals. This joint effort between government and corporations is essential for addressing complex global challenges, such as climate change and biodiversity loss.

Ultimately, when businesses and governments collaborate on sustainability, the impact can be transformative. By working together, these entities can create a more sustainable future and contribute to the well-being of both society and the environment.

Innovations in Renewable Energy and Corporate Responsibility

The energy sector plays a pivotal role in the transition to a sustainable future. As part of their corporate responsibility, businesses are increasingly investing in renewable energy sources such as solar, wind, and geothermal power. Renewable energy not only helps companies reduce their carbon emissions but also positions them as leaders in environmental sustainability (Brown, 2018).

Companies in various industries, from tech giants like Google and Apple to manufacturers like General Electric, have made substantial investments in renewable energy to power their operations. By transitioning to renewable energy sources, these companies not only reduce their environmental impact but also contribute to the growth of the green energy sector. This transition is essential for achieving global sustainability goals, including the reduction of greenhouse gas emissions and the promotion of clean energy (International Energy Agency, 2020).

In addition to direct investments in renewable energy, companies are increasingly participating in renewable energy purchase agreements (REAs), which allow businesses to purchase renewable energy directly from energy providers. These agreements help businesses secure a reliable and cost-effective source of clean energy while driving demand for renewable energy projects. Moreover, companies that invest in renewable energy can benefit from cost savings in the long run due to lower energy expenses.

By embracing renewable energy, companies not only align with global sustainability efforts but also demonstrate a long-term commitment to environmental responsibility. This shift represents a crucial step in the global transition to green power.

Corporate Transparency and Environmental Impact

Corporate transparency in environmental impact reporting is increasingly becoming a requirement for businesses seeking to demonstrate their commitment to sustainability. Transparency involves publicly sharing information about a company's environmental performance, including energy use, emissions, waste management, and resource consumption. By being transparent about their environmental impact, companies foster trust with consumers, investors, and regulators, which can positively influence their reputation and profitability (Eccles et al., 2012).

Environmental disclosure is typically done through sustainability reports, carbon footprint assessments, and third-party certifications. Companies such as Patagonia and Ben & Jerry's are known for their transparency in sharing their sustainability initiatives and environmental impact data. These companies not only disclose the challenges they face in achieving sustainability but also set clear goals for improvement, which aligns with their corporate responsibility strategy (Eccles et al., 2012).

Governments and regulatory bodies also play a role in encouraging corporate transparency through policies that mandate environmental reporting. For example, the European Union requires companies to disclose information about their environmental practices, while the Global Reporting Initiative (GRI) provides a framework for sustainability reporting that many companies adopt.

As corporate transparency continues to grow, businesses are increasingly held accountable for their actions, which helps drive continuous improvement in sustainability practices.

Ultimately, corporate transparency is essential for building credibility and trust, ensuring that companies are genuinely contributing to environmental preservation, and enabling stakeholders to make informed decisions about their investments.

The Role of Marketing in Promoting Sustainable Business Practices

Marketing plays a crucial role in promoting corporate responsibility by educating consumers and encouraging sustainable purchasing decisions. Green marketing involves promoting products and services that have a minimal environmental impact, thereby influencing consumer choices and behavior (Peattie, 2010). This marketing strategy has gained significant traction as environmental concerns have become more central to consumer decision-making.

One of the key elements of green marketing is communicating the environmental benefits of products or services. Companies use various platforms, including advertisements, social media, and product labeling, to showcase their environmental efforts. For instance, companies like Seventh Generation and The Body Shop have used green marketing to emphasize the sustainability of their products, such as using organic ingredients or eco-friendly packaging (Peattie, 2010). By highlighting their environmental commitments, companies build brand loyalty and attract consumers who prioritize sustainability.

Greenwashing, however, is a concern in the realm of green marketing. Greenwashing occurs when companies make misleading claims about their environmental practices to capitalize on the growing demand for sustainable products. To counter this, consumers and regulatory bodies are becoming more vigilant in verifying the authenticity of environmental claims. Third-party certifications such as the Energy Star label or Fair-Trade certification provide consumers with reliable information about a product's environmental credentials.

As consumer demand for sustainable products continues to grow, green marketing will play an essential role in promoting environmentally responsible business practices and driving positive change.

Corporate Social Responsibility and Stakeholder Engagement

Corporate Social Responsibility (CSR) has evolved to include not only environmental sustainability but also social and economic considerations. Companies that engage in CSR recognize the need to contribute positively to society while balancing profit-making activities. By addressing the needs and concerns of stakeholders—such as employees, consumers, local communities, and investors—companies ensure that their business practices align with broader societal goals (Carroll, 1999).

Stakeholder engagement is central to CSR because it enables businesses to understand and respond to the expectations of different groups. For instance, companies in the energy sector may engage with local communities to address concerns about the environmental impact of their operations. In the retail sector, brands may engage with consumers to understand their preferences for sustainable products and ethical practices.

Effective CSR strategies require transparent communication and active collaboration with stakeholders. Companies must not only engage in dialogue but also take concrete actions that demonstrate their commitment to sustainability. For example, many businesses are adopting fair labor practices, ensuring that their supply chains are free of child labor, and supporting local communities through charitable initiatives (Carroll, 1999). These efforts create goodwill, strengthen brand reputation, and contribute to long-term business success. As CSR becomes more integrated into corporate strategies, businesses are increasingly recognizing that addressing social, environmental, and economic challenges benefits both society and their bottom line.

Corporate Responsibility and Global Supply Chains

Global supply chains have become essential for businesses operating in a globalized economy, but they also pose significant challenges in terms of environmental and social responsibility. Companies today are increasingly focusing on ensuring that their supply chains are sustainable, ethical, and transparent. Responsible supply chain management involves working with suppliers to reduce the environmental impact of operations, improve labor conditions, and promote fair trade practices (Seuring & Müller, 2008).

For instance, major companies like Nike and Apple have taken steps to improve the sustainability of their supply chains by working with suppliers to reduce carbon emissions, use sustainable materials, and ensure fair labor practices. Additionally, global corporations are using tools such as the Global Reporting Initiative (GRI) and the Supplier Ethical Data Exchange (SEDEX) to monitor their supply chain activities and ensure compliance with sustainability standards.

By adopting sustainable supply chain practices, companies can improve their environmental performance, reduce costs, and build stronger relationships with stakeholders. Moreover, responsible supply chain management helps companies mitigate risks associated with supply chain disruptions, such as those caused by environmental disasters or labor violations. Ultimately, businesses that prioritize sustainability in their global supply chains not only reduce their environmental footprint but also enhance their reputation and contribute to a more sustainable global economy.

The Future of Corporate Responsibility and Environmental Sustainability

As the world continues to grapple with the environmental challenges posed by climate change, resource depletion, and pollution, the role of businesses in driving sustainability has never been more critical. Corporate responsibility is evolving to meet the challenges of the 21st century, with companies taking a more active role in reducing their environmental impact and contributing to global sustainability efforts.

The future of corporate responsibility will likely see an increasing emphasis on integrating sustainability into core business strategies. Companies will continue to innovate and adopt new technologies that minimize waste, reduce emissions, and promote the use of renewable resources. Additionally, businesses will increasingly focus on circular economy models, where products are designed for reuse, repair, and recycling, rather than disposal (Geissdoerfer et al., 2017).

Moreover, as consumer demand for sustainable products rises, companies that prioritize environmental responsibility will gain a competitive advantage. Corporate responsibility will no longer be seen as a separate initiative but as a fundamental component of business operations. In this way, businesses will play a pivotal role in achieving the United Nations Sustainable Development Goals (SDGs), which aim to create a more sustainable and equitable world by 2030.

Ultimately, the future of corporate responsibility and environmental sustainability will require businesses to innovate, collaborate, and lead by example. As more companies embrace sustainability, they will collectively contribute to a greener, more sustainable future.

Chapter 15: Community Action: Grassroots movements

Introduction to Community Action and Grassroots Movements

Community action refers to collective efforts undertaken by a group of individuals to address specific issues or improve their local environment. Grassroots movements, a form of community action, are typically organized at the local level, relying on the energy, commitment, and collaboration of ordinary citizens. These movements play a crucial role in social change by empowering communities to take control of their own futures, challenge existing power structures, and advocate for their rights (Hochschild, 2018).

Grassroots movements are often driven by community members who share common interests, concerns, and goals. These movements are characterized by their decentralized, bottom-up approach, with decision-making processes involving a wide range of stakeholders. Unlike top-down initiatives led by government agencies or corporations, grassroots movements prioritize direct participation and inclusivity (Barkan, 2018).

In the face of environmental degradation, economic inequality, and social injustice, grassroots movements have emerged as powerful tools for enacting change. From environmental justice campaigns to movements advocating for workers' rights, grassroots initiatives have been at the forefront of addressing systemic issues and pushing for policy reforms. These movements demonstrate the potential of ordinary citizens to create lasting, meaningful change in their communities.

This paper will explore the significance of community action, focusing on grassroots movements, their impact, challenges, and potential for driving societal transformation. By examining case studies and drawing on research, the paper will highlight how grassroots movements have shaped the social, political, and environmental landscape globally.

Historical Context of Grassroots Movements

Grassroots movements have a rich history, emerging out of the need for marginalized communities to have a voice in the decisions that affect their lives. One of the most notable examples of grassroots activism is the Civil Rights Movement in the United States, which began as a local initiative but expanded into a nationwide struggle for racial equality (McAdam, 2019). This movement, like many others, was rooted in community action, where ordinary citizens organized protests, sit-ins, and marches to demand civil rights.

Historically, grassroots movements have been an essential response to systemic oppression and injustice. In the early 20th century, labor movements gained momentum as workers sought better wages, working conditions, and the right to unionize. These movements were initiated and led by working-class people, not political elites, and were fundamental in shaping labor rights today (Tarrow, 2018). Similarly, the environmental movement of the 1960s and 1970s arose in response to widespread pollution, hazardous working conditions, and the degradation of public health. The growth of these movements was fueled by individuals who were directly impacted by environmental issues, particularly in local communities.

The historical trajectory of grassroots movements underscores their ability to challenge existing power structures and create tangible, lasting change. These movements have often led to legislative reforms, shifts in public opinion, and cultural changes that have transformed society. Thus, understanding the historical foundations of grassroots movements is key to recognizing their ongoing importance in social and political activism today.

Key Characteristics of Grassroots Movements

Grassroots movements share several defining characteristics that distinguish them from traditional forms of activism. First and foremost, these movements are typically initiated and led by local individuals rather than large, centralized organizations or political figures (Barkan, 2018). This bottom-up approach allows for greater inclusion and representation of the community's needs and concerns. Grassroots movements often form in response to immediate issues that directly affect the community, such as environmental degradation, unemployment, or inadequate public services (Hochschild, 2018).

Another key characteristic of grassroots movements is their emphasis on direct action. Activists engage in activities such as protests, strikes, boycotts, and petitions to demand change from local governments or corporations. These actions are designed to draw attention to the issue at hand and to put pressure on decision-makers. Moreover, grassroots movements typically advocate for systemic change rather than piecemeal reforms, seeking to address the root causes of social, economic, or environmental problems (Tarrow, 2018).

Finally, grassroots movements are known for their adaptability and ability to mobilize quickly. Because they are often decentralized and not reliant on formal organizational structures, these movements can respond rapidly to changing circumstances. Social media and digital platforms have further enhanced the reach and speed of grassroots organizing, allowing movements to spread globally with minimal resources.

In sum, the defining traits of grassroots movements—local leadership, direct action, and adaptability—make them powerful agents of change in contemporary society.

The Role of Social Media in Grassroots Movements

In the digital age, social media has become an invaluable tool for grassroots movements, enabling them to reach wider audiences and mobilize support more effectively. Platforms such as Twitter, Facebook, Instagram, and TikTok have provided a space for individuals and groups to organize, share information, and advocate for change in real-time (McAdam, 2019). Social media allows activists to bypass traditional media channels, which may be biased or inaccessible, and communicate directly with the public.

One of the most significant ways social media has impacted grassroots movements is by facilitating rapid mobilization. For example, during the Arab Spring protests in 2010 and 2011, social media platforms played a critical role in organizing demonstrations, spreading information, and fostering international solidarity (Tufekci, 2017). Protestors used platforms like Twitter to organize protests, share live updates, and call attention to human rights violations, helping to galvanize global support for their cause.

Furthermore, social media has made it easier for marginalized voices to be heard. Activists who may have once struggled to gain visibility can now use social media to amplify their message and reach a global audience. The #MeToo movement, for instance, gained widespread attention through social media, empowering individuals to share their experiences with sexual harassment and catalyzing a global conversation about gender equality (Tufekci, 2017).

While social media has enhanced grassroots organizing, it is not without its challenges. Activists must contend with misinformation, censorship, and surveillance, which can undermine the effectiveness of their efforts. Nonetheless, the role of social media in grassroots movements remains pivotal in modern activism.

Case Study: The Civil Rights Movement

The Civil Rights Movement (CRM) in the United States serves as one of the most prominent examples of a grassroots movement achieving significant social change. Born out of the need to end racial segregation and secure equal rights for African Americans, the movement emerged in the mid-20th century, propelled by local efforts and community organizing. Figures such as Rosa Parks, Martin Luther King Jr., and others became symbols of the broader movement, but the foundation of the CRM lay in the efforts of countless individuals across the country who advocated for civil rights.

One key aspect of the Civil Rights Movement was its use of nonviolent resistance. Activists engaged in sit-ins, marches, and boycotts to protest segregationist policies and demand equal rights. One of the most famous examples of grassroots action during this period was the 1963 March on Washington, which brought together more than 250,000 people from various walks of life to demand racial equality (McAdam, 2019). This massive mobilization demonstrated the power of grassroots activism in bringing attention to an issue that had been ignored for decades.

The success of the Civil Rights Movement was not solely due to the efforts of high-profile leaders; rather, it was the result of coordinated grassroots actions across the nation. By organizing in local communities and engaging in direct action, activists were able to pressure lawmakers and policymakers to pass landmark legislation such as the Civil Rights Act of 1964 and the Voting Rights Act of 1965.

Case Study: The Environmental Justice Movement

The environmental justice movement is a powerful example of grassroots action aimed at addressing environmental issues, particularly those affecting marginalized communities. Emerging in the United States in the 1980s, this movement sought to highlight the disproportionate impact of environmental hazards on low-income and minority populations (Bullard, 2018). Environmental justice advocates argue that environmental policies often neglect the needs of vulnerable communities, who face higher risks of exposure to pollution, toxic waste, and climate-related disasters.

One of the most notable cases that sparked the environmental justice movement was the 1982 protests in Warren County, North Carolina. Residents, mostly African American, organized against the siting of a hazardous waste landfill in their community. The protest, although ultimately unsuccessful in halting the landfill's construction, marked a turning point in how environmental issues were viewed and addressed (Bullard, 2018). It demonstrated the interconnectedness of environmental degradation and social inequality, and helped to build a coalition of grassroots groups advocating for environmental justice.

The movement has since expanded globally, with grassroots organizations working to ensure that environmental policies are equitable and just for all people, regardless of their socio-economic status or race. Today, environmental justice advocates continue to fight for stronger regulatory protections, better community engagement in decision-making processes, and improved environmental health in underserved areas. Their work highlights the need for inclusive solutions to global environmental challenges.

The Role of Education in Grassroots Movements

Education plays a central role in the success of grassroots movements. Through education, activists are able to raise awareness, mobilize individuals, and empower communities to take action. One key component of grassroots movements is the ability to inform individuals about issues that directly affect their lives, whether they relate to social justice, environmental protection, or public health.

In the context of environmental activism, for example, education is often the first step in catalyzing change. Environmental organizations and grassroots groups work to educate communities about the risks of pollution, climate change, and habitat destruction.

By providing accurate information, they help individuals understand the scale of the problem and the actions they can take to mitigate its effects. Education is also crucial for fostering a sense of shared responsibility within communities, allowing them to work collectively toward solutions (Schroeder & McCarthy, 2020).

Moreover, education helps to build leadership within grassroots movements. By equipping community members with the tools and knowledge necessary for effective advocacy, these movements can sustain themselves over time. Workshops, seminars, and training programs are often used to teach participants about the principles of organizing, the importance of strategic communication, and the legal and policy frameworks that govern their areas of concern (Schroeder & McCarthy, 2020). In this way, education is not only a means of raising awareness but also a catalyst for long-term social and environmental change.

Case Study: The Women's Suffrage Movement

The women's suffrage movement, which led to women gaining the right to vote in many parts of the world, is another key example of a grassroots initiative. This movement, which gained momentum in the late 19th and early 20th centuries, was driven by the tireless efforts of women who fought for gender equality. Activists such as Susan B. Anthony, Elizabeth Cady Stanton, and Emmeline Pankhurst were instrumental in organizing protests, petitions, and civil disobedience campaigns to demand suffrage rights for women (Tosh, 2018).

Despite facing significant opposition, suffragists mobilized thousands of women at the local, national, and international levels. Through grassroots activism, women were able to raise awareness about the injustices they faced and challenge deeply ingrained social norms that limited their participation in public life. One of the most notable successes of the women's suffrage movement was the passage of the 19th Amendment in the United States in 1920, which granted women the right to vote (Tosh, 2018). The suffrage movement also played a key role in advancing other women's rights issues, including access to education, employment, and reproductive rights.

The women's suffrage movement is a testament to the power of grassroots activism in bringing about meaningful social change. It highlights the importance of persistent organizing, coalition-building, and strategic mobilization in achieving long-term goals. Moreover, the legacy of the suffrage movement continues to inspire contemporary feminist movements around the world.

Building Solidarity in Grassroots Movements

A key factor in the success of grassroots movements is the ability to build solidarity among diverse groups. Solidarity refers to the sense of unity and collective action that is essential for mobilizing individuals around a common cause. In grassroots movements, solidarity is particularly important because it brings together people with shared interests, regardless of their backgrounds or individual circumstances.

One of the most powerful ways grassroots movements build solidarity is by focusing on the shared values and goals of the community. Whether addressing environmental concerns, social justice, or economic inequality, these movements often emphasize the importance of collective action in achieving a better future. By fostering a sense of common purpose, grassroots movements create a foundation for cooperation and mutual support among participants (Hochschild, 2018).

Solidarity is also built through direct engagement and communication. Grassroots movements often rely on face-to-face interactions, community meetings, and local organizing to build trust and understanding. Through these interactions, activists and community members can share their experiences, discuss strategies, and work together to identify solutions. Solidarity is not just about agreeing on a set of goals—it is about forging relationships and creating a sense of shared responsibility for the movement's success.

The ability to build solidarity is particularly crucial in the face of opposition. Grassroots movements often encounter resistance from powerful institutions, including governments and corporations. By maintaining a sense of unity and collective strength, activists can overcome challenges and continue to push for change.

Challenges Faced by Grassroots Movements

While grassroots movements have proven to be powerful agents of social and environmental change, they also face numerous challenges. One of the most significant obstacles is limited access to resources. Unlike large organizations or political entities, grassroots movements typically operate on minimal budgets, which can constrain their ability to carry out campaigns, conduct research, and engage with policymakers (Barkan, 2018). Additionally, the lack of financial resources can hinder efforts to organize large-scale demonstrations, hire staff, or purchase materials for educational outreach.

Another challenge is the difficulty in maintaining momentum over time. Grassroots movements often rely on volunteer labor and passionate individuals who may burn out from the intensity of their work. Activists must constantly find ways to sustain engagement and recruit new participants. This challenge is particularly pronounced in movements focused on long-term goals, such as combating climate change or fighting for racial justice (Schroeder & McCarthy, 2020).

Moreover, grassroots movements often face opposition from powerful entities such as governments, corporations, or the media. These actors may use legal, political, or economic strategies to undermine the legitimacy of the movement and its goals. In some cases, governments have resorted to censorship or surveillance to monitor and control activist activities. The ability of grassroots movements to navigate these challenges and persevere is a testament to the resilience and determination of the individuals involved.

The Importance of Local Leadership in Grassroots Movements

In grassroots movements, local leadership is critical to the success and sustainability of activism. Local leaders play a pivotal role in connecting the movement to the community, organizing actions, and maintaining momentum. These leaders are often from the communities they serve and have a deep understanding of local issues and concerns. Their knowledge of the cultural, social, and economic landscape enables them to engage people effectively and address the unique challenges faced by the community (Mann, 2019).

One of the key strengths of local leadership is the ability to create a sense of ownership within the community. When people see leaders who share their experiences and struggles, they are more likely to get involved and participate in the movement. Local leaders also serve as the primary point of contact for residents and are essential in organizing meetings, protests, and advocacy campaigns (Mann, 2019). Through their leadership, they help to forge a sense of unity and shared purpose among community members.

Moreover, local leaders often act as liaisons between the grassroots movement and external actors, such as local governments, non-profit organizations, or media outlets. Their role in building these relationships can help secure resources, bring attention to the cause, and influence policy decisions. Local leadership is especially important in movements that focus on specific local issues, such as environmental justice, community development, or housing rights.

Grassroots Movements in the Digital Age

The rise of digital technology has transformed grassroots movements in unprecedented ways. The internet and social media have provided activists with powerful tools to reach a global audience, mobilize support, and coordinate actions. Digital platforms like Facebook, Twitter, Instagram, and YouTube allow grassroots organizations to amplify their messages and connect with individuals around the world, bypassing traditional media channels that may have previously marginalized their voices (McCarthy & Zald, 2020).

Social media, in particular, has become an essential tool for organizing protests, spreading awareness, and creating viral campaigns. Hashtags, online petitions, and live streams of demonstrations can generate significant attention and support. For example, the #BlackLivesMatter movement gained global traction through Twitter, where individuals shared their experiences and advocated for racial justice. Similarly, environmental movements like Fridays for Future have utilized social media to raise awareness about climate change and engage millions of young people worldwide (McCarthy & Zald, 2020).

While the digital age has opened up new avenues for grassroots movements, it has also introduced challenges. The spread of misinformation, online harassment, and the potential for censorship by governments or corporations can hinder the effectiveness of digital activism. Despite these challenges, the internet remains a vital tool for modern grassroots movements, allowing them to expand their reach and create impactful change on a global scale.

Global Grassroots Movements: Learning from One Another

While grassroots movements often begin at the local level, they can quickly gain national and even international support. One of the key advantages of grassroots movements is the ability to share experiences and learn from one another. Movements in one country can inspire similar efforts in other parts of the world, leading to a shared global network of activists and organizations working toward common goals (Tarrow, 2018).

For example, the global environmental movement has seen collaboration between grassroots organizations from different regions of the world. In the 1990s, activists from the Global South, particularly Latin America and Africa, began to collaborate with environmental groups in the Global North to address environmental issues such as deforestation, water pollution, and biodiversity loss. This collaboration helped to elevate the voices of marginalized communities and increase international awareness about the impacts of environmental degradation on vulnerable populations (Tarrow, 2018).

Similarly, the international women's rights movement has benefited from the exchange of strategies and resources among grassroots groups worldwide. From the fight for reproductive rights in the United States to the struggle for gender equality in India, women's rights organizations have supported each other in advancing social and political change. The ability to learn from successful movements and share resources has enhanced the effectiveness of grassroots initiatives and helped to create a more unified global movement for justice and equality.

The Role of Art and Culture in Grassroots Movements

Art and culture have long played a vital role in grassroots movements, providing a means of expression, protest, and resistance. Creative forms of expression, such as music, visual art, theater, and literature, help communicate the messages of grassroots movements in ways that resonate with people on an emotional level. These cultural expressions can also serve as tools for building solidarity and inspiring action within communities.

For example, protest songs and music have been integral to movements for social justice. During the civil rights movement in the United States, songs like "We Shall Overcome" became anthems of resistance and unity. Similarly, artists in South Africa used music and visual art to protest apartheid and raise awareness about the oppression faced by black South Africans (Hutchinson, 2020).

In addition to music, visual art can also be a powerful tool for communicating messages of protest. Street art, murals, and graffiti have been used in movements around the world to challenge political regimes, raise awareness of social issues, and inspire action. For example, murals in the United States and Latin America often depict the struggles of marginalized communities and serve as symbols of resistance and resilience (Hutchinson, 2020).

Art and culture not only help to communicate the goals of grassroots movements, but they also create a sense of identity and belonging for those involved. By celebrating the cultural richness of communities and elevating marginalized voices, art serves as a powerful tool for change.

The Future of Grassroots Movements

As the world faces an increasing number of challenges—ranging from climate change and inequality to political instability and human rights abuses—grassroots movements will continue to play a vital role in driving change. The future of grassroots movements looks promising, as more people around the world recognize the power of collective action to address social and environmental issues. With advances in technology, greater access to information, and the growing interconnectedness of global communities, grassroots movements have the potential to bring about transformative change on a scale never before seen.

One of the key trends shaping the future of grassroots movements is the increased focus on intersectionality. Activists are recognizing that social issues, such as gender inequality, racial discrimination, and environmental degradation, are interconnected and must be addressed together. This understanding is leading to more inclusive and holistic approaches to activism, as well as a greater emphasis on collaboration across movements (Collins, 2020).

Additionally, the continued growth of digital platforms is likely to enhance the reach and impact of grassroots movements. While challenges such as misinformation and online censorship remain, the ability to mobilize individuals, share information, and create online communities has opened up new opportunities for grassroots organizing. The power of social media to spark global movements—such as the Arab Spring or the global climate strikes—is likely to continue playing a crucial role in activism in the future.

Grassroots Movements and Their Political Impact

Grassroots movements have historically had a significant impact on political landscapes, often pushing for changes in laws, policies, and social norms. These movements bring issues to the forefront of public consciousness and demand that governments take action. By organizing protests, lobbying policymakers, and engaging in direct action, grassroots movements can create political pressure that forces government officials to respond to the demands of the people.

One of the key examples of political impact is the Civil Rights Movement in the United States. Through direct action, legal challenges, and nationwide protests, the movement succeeded in securing landmark legislation such as the Civil Rights Act of 1964 and the Voting Rights Act of 1965 (Garrow, 2021). Similarly, grassroots movements in the global South have influenced the political trajectory of several countries by advocating for the rights of indigenous populations, workers, and environmental activists. In many cases, these movements have led to significant reforms in land rights, labor laws, and environmental protections (Tarrow, 2018).

Today, grassroots movements continue to shape the political landscape. The #MeToo movement, for example, has prompted widespread conversations about sexual harassment and assault, leading to policy changes in workplaces, schools, and governments around the world. By demanding accountability and equality, grassroots movements challenge entrenched systems of power and push for political reform.

The future of grassroots movements in the political arena depends on their ability to adapt to changing political climates, build coalitions, and continue to press for social justice.

The Role of Media in Grassroots Movements

Media plays an essential role in the success of grassroots movements by amplifying their messages, mobilizing support, and keeping the public informed about their actions. Traditional media outlets like newspapers, radio, and television have long been instrumental in bringing attention to social and environmental causes. However, with the rise of digital media, grassroots movements have gained new opportunities to share their messages and reach wider audiences (Castells, 2019).

In the digital age, social media platforms such as Twitter, Facebook, and Instagram have become vital tools for organizing, communicating, and rallying support. Activists use these platforms to share stories, spread information, and call for action. Hashtags like #BlackLivesMatter and #FridaysForFuture have become global movements, engaging millions of people worldwide in conversations about race, climate change, and social justice. Social media enables movements to break free from traditional media gatekeepers, allowing grassroots activists to directly engage with their audiences and bypass mainstream media narratives (McChesney, 2018).

In addition to social media, citizen journalism has emerged as an influential form of media that helps amplify the voices of grassroots movements. Through blogs, podcasts, and video content, activists can document their experiences, share firsthand accounts of protests, and report on issues that may not receive coverage from mainstream outlets (Castells, 2019). Citizen journalism plays a critical role in empowering grassroots movements, giving activists the tools they need to communicate their stories and challenge established narratives.

The Role of Non-Governmental Organizations (NGOs) in Grassroots Movements

Non-governmental organizations (NGOs) play a crucial role in supporting grassroots movements by providing resources, advocacy, and expertise. NGOs often act as intermediaries between grassroots organizations and international actors, such as governments, corporations, and donors. By offering logistical support, financial aid, and access to networks, NGOs can help grassroots movements scale up their efforts and amplify their impact (Korten, 2017).

Many NGOs focus on issues such as human rights, environmental protection, and social justice, aligning their missions with those of grassroots movements. For example, organizations like Greenpeace and the World Wildlife Fund (WWF) work closely with local communities to address environmental challenges and promote sustainable development. These NGOs often collaborate with grassroots organizations to conduct research, engage in policy advocacy, and organize public campaigns (Keck & Sikkink, 2018).

While NGOs can provide valuable support, it is essential that they maintain a collaborative and respectful relationship with grassroots movements. NGOs must recognize the importance of local leadership and ensure that their activities align with the priorities of the communities they serve. In some cases, NGOs have been criticized for imposing external agendas or working in ways that undermine the autonomy of grassroots groups. Therefore, a balanced approach that respects local knowledge and leadership is necessary for successful collaboration.

The Power of Youth in Grassroots Movements

The involvement of young people in grassroots movements has become increasingly prominent in recent years. From climate change protests to campaigns for social justice, youth-led movements have made significant strides in raising awareness about global issues and pushing for change. The power of youth lies in their passion, energy, and ability to challenge the status quo.

One of the most notable examples of youth-driven grassroots movements is the Fridays for Future campaign, which was founded by Swedish climate activist Greta Thunberg. Thunberg's school strike for climate, which began in 2018, has sparked a global movement that has mobilized millions of young people around the world to demand action on climate change. The movement emphasizes the need for immediate policy action to address the climate crisis and has called on governments to take stronger measures to reduce carbon emissions (Thunberg, 2019).

Youth-led movements often focus on issues that directly affect young people, such as education, job opportunities, and environmental sustainability. These movements challenge the traditional power structures and advocate for a more inclusive and sustainable future. Through their activism, young people are not only raising awareness but also reshaping the political and social landscape, demanding that future generations inherit a healthier, fairer world.

The rise of youth activism signals a shift in how grassroots movements operate. Young people are using technology, social media, and global networks to organize and make their voices heard. Their ability to engage and mobilize on a global scale is reshaping the future of activism.

Conclusion: The Future of Grassroots Movements

Grassroots movements are essential to driving social and environmental change. Through collective action, they empower individuals, raise awareness, and challenge systems of power. Over time, these movements have proven to be incredibly effective at influencing policy, reshaping public attitudes, and bringing attention to critical issues such as climate change, racial justice, and economic inequality.

The future of grassroots movements lies in their ability to adapt to the ever-changing political and technological landscape. As digital tools and social media continue to shape activism, grassroots movements will likely become more interconnected and global in nature. The power of local leadership, collaboration, and the engagement of diverse communities will remain at the heart of successful movements.

For grassroots movements to continue their impact, they must remain focused on intersectionality, building solidarity, and empowering marginalized voices. As the world faces increasing challenges, the role of grassroots movements will become even more crucial in shaping a more equitable and sustainable future.

The resilience, creativity, and determination of grassroots activists ensure that change is not only possible but inevitable. Whether addressing local issues or global crises, grassroots movements will continue to be a force for positive transformation.

Chapter 16: Education and Awareness: Spreading knowledge

Introduction to Education and Awareness

Education and awareness are powerful tools for fostering social, environmental, and economic change. By spreading knowledge, individuals are empowered to make informed decisions, challenge injustices, and contribute to collective well-being. Education, particularly in the areas of environmental sustainability, health, and social justice, helps individuals understand the complexities of the world and their role in shaping a better future (UNESCO, 2020).

Awareness programs aim to inform people about important issues, such as climate change, pollution, and human rights. These efforts can take many forms, from public campaigns and community discussions to educational programs in schools and universities. By raising awareness, organizations and activists inspire individuals to take action and become agents of change in their communities. Education fosters critical thinking, enabling people to question the status quo and seek alternative solutions to societal challenges (Smith, 2019).

For education to be effective, it must be accessible and relevant to the target audience. Different methods, such as interactive workshops, media campaigns, and digital resources, can be employed to reach diverse populations. The goal is not only to inform but also to inspire long-lasting changes in behavior and attitudes. By ensuring that education and awareness initiatives are inclusive and engaging, societies can build a more informed and proactive citizenry capable of addressing pressing global challenges (Miller, 2018).

The Role of Education in Social Change

Education is a fundamental driver of social change, shaping the values, norms, and behaviors of individuals and societies. When people are educated, they are more likely to engage in social and political activities that promote justice, equity, and sustainability. Education empowers people to challenge existing power structures, advocate for policy change, and build more inclusive societies (Freire, 1970).

For instance, the global push for gender equality has been significantly influenced by educational initiatives that raise awareness about women's rights. By educating people about gender equality, movements like #MeToo have gained momentum and created significant societal shifts. Similarly, education has played a crucial role in environmental justice movements, where people are educated about the harmful impacts of pollution, deforestation, and climate change. This awareness prompts individuals to take actions such as reducing waste, supporting green policies, and advocating for the protection of natural resources (Wallerstein, 2019).

Furthermore, education in the form of advocacy training, public campaigns, and policy workshops helps individuals and groups gain the skills and knowledge needed to engage with decision-makers. This level of education not only empowers individuals but also creates a ripple effect, where the educated individual becomes a resource for others, spreading awareness and facilitating further education on pressing social issues (Freire, 1970).

Raising Awareness About Climate Change

Climate change is one of the most pressing issues of our time, and education is crucial in raising awareness about its causes, impacts, and solutions. Climate change education provides individuals with the knowledge needed to understand the science behind climate change, the risks it poses, and the ways in which they can reduce their carbon footprint. By educating people about climate change, societies can take collective action to mitigate its effects and build more resilient communities.

Public awareness campaigns, school programs, and grassroots initiatives are all essential tools in climate change education. Through these efforts, people are made aware of how human activities, such as burning fossil fuels, deforestation, and industrial pollution, contribute to global warming. In addition, educational initiatives often focus on providing solutions, such as renewable energy sources, energy-efficient practices, and sustainable agricultural methods (Hulme, 2018).

Awareness campaigns are particularly effective in fostering collective action. When individuals understand the magnitude of climate change and its potential impacts on future generations, they are more likely to support policies that reduce emissions, protect biodiversity, and promote sustainable living. Education is the foundation for building a more environmentally conscious society that is motivated to act on climate change and support sustainable practices.

The Role of Media in Spreading Knowledge

Media plays a significant role in spreading knowledge and raising awareness on various issues, including environmental, social, and health-related matters. News outlets, documentaries, podcasts, and online platforms are powerful tools for informing the public about critical issues. Through media, complex topics are made accessible, and individuals are given the opportunity to engage with new ideas and perspectives (Anderson, 2019).

Media campaigns are often used to educate the public on pressing matters such as pollution, climate change, and the importance of conservation. For example, environmental organizations use documentaries and social media platforms to showcase the beauty of natural ecosystems and the threats they face from human activity. These visual stories can be compelling, evoking emotional responses and motivating viewers to take action (Barker, 2020).

In addition to educating the public, media also plays a crucial role in shaping public opinion. The way in which media portrays an issue can influence how the public perceives its importance and urgency. For example, media coverage of climate change has increased public awareness and led to greater demands for political action. Therefore, media serves not only as a tool for spreading knowledge but also as a means of fostering societal change through the dissemination of critical information (Anderson, 2019).

Education for Sustainable Development

Education for Sustainable Development (ESD) is an educational approach that integrates the principles of sustainability into teaching and learning. ESD aims to equip individuals with the knowledge, skills, and values needed to contribute to sustainable development and address environmental challenges. This approach emphasizes the importance of educating people about sustainability, climate change, and the conservation of natural resources (UNESCO, 2017).

Incorporating sustainability into education helps individuals understand the interconnectedness of social, economic, and environmental systems. It encourages students to think critically about their role in creating a sustainable future and empowers them to make informed decisions. For example, students can learn about sustainable agriculture, renewable energy sources, and waste management practices that reduce environmental impact. These lessons not only benefit the students but also influence their families and communities as they share their knowledge (UNESCO, 2017).

ESD also promotes the values of equity, justice, and peace, recognizing that sustainable development must address social and economic inequalities. By educating individuals about the importance of these values, ESD fosters a sense of responsibility and social consciousness. As a result, ESD contributes to the formation of a global citizenry that is committed to protecting the environment and promoting social well-being.

The Importance of Environmental Education in Schools

Environmental education in schools is essential for equipping the younger generation with the knowledge and skills required to tackle pressing global challenges, such as climate change and environmental degradation. By integrating environmental issues into the curriculum, students are given the tools to understand the impact of human activity on the planet and are encouraged to make sustainable choices. Environmental education fosters a sense of responsibility and empowers students to become stewards of the environment (Sterling, 2017).

Schools can incorporate environmental topics into a wide range of subjects, from science to geography and even ethics. Lessons can cover topics like ecosystems, biodiversity, climate change, renewable energy, and sustainable farming. Field trips, nature-based projects, and hands-on activities help students connect theory with practice and develop problem-solving skills. When students actively engage with environmental issues, they are more likely to internalize the importance of sustainability and apply this knowledge in their everyday lives.

Moreover, environmental education helps cultivate critical thinking and promotes an understanding of the interconnections between society, the economy, and the environment. It encourages students to question unsustainable practices, advocate for policy changes, and seek innovative solutions. By fostering a generation of environmentally conscious citizens, environmental education in schools contributes to the long-term goal of creating a sustainable future for all (Sterling, 2017).

Global Initiatives to Raise Awareness

Globally, numerous initiatives are focused on raising awareness about critical environmental and social issues. The United Nations, through its Sustainable Development Goals (SDGs), has played a central role in promoting education for sustainable development and spreading knowledge about issues such as climate action, clean water, and biodiversity conservation. These global initiatives provide a framework for governments, organizations, and communities to work together in raising awareness and implementing solutions to global challenges (United Nations, 2020).

Campaigns like Earth Day, World Environment Day, and the 350.org climate movement have mobilized millions of people around the world to take action for the environment. These global events draw attention to specific environmental issues and inspire people to participate in activities like tree planting, waste reduction, and advocacy for policy changes. Through social media and other communication platforms, awareness of these events has reached an unprecedented scale, creating a global network of individuals united by a common cause.

Educational resources, such as documentaries, public service announcements, and online platforms, also contribute to these global efforts. Initiatives that focus on spreading knowledge about the impacts of climate change and environmental degradation encourage individuals to adopt sustainable behaviors and advocate for systemic change. By engaging in these global movements, individuals not only increase their awareness but also become active participants in the fight for a more sustainable world (United Nations, 2020).

Using Technology to Spread Awareness

Technology plays a crucial role in spreading knowledge and raising awareness about global issues. The internet, social media platforms, mobile apps, and other digital tools have made it easier to share information and engage with diverse audiences. These technological advancements allow educational content to reach millions of people in real-time, making it possible for individuals and organizations to spread awareness about social, environmental, and political issues on a global scale (Anderson, 2019).

Social media platforms like Twitter, Facebook, and Instagram have become important tools for raising awareness about climate change, biodiversity loss, and human rights violations. These platforms allow individuals and organizations to share articles, videos, and infographics that educate the public and spark conversations. For example, the viral #FridaysForFuture movement, initiated by Greta Thunberg, uses social media to raise awareness about climate change and mobilize people to take action.

Moreover, technology enables the creation of interactive and engaging educational content. Online courses, webinars, and podcasts provide opportunities for people to learn about environmental issues at their own pace. Educational websites and mobile apps also offer practical tips for adopting sustainable practices, such as reducing energy consumption, minimizing waste, and supporting eco-friendly products. As technology continues to evolve, it will undoubtedly play an even greater role in raising awareness and inspiring individuals to take action (Anderson, 2019).

Environmental Advocacy and Activism

Environmental advocacy and activism are integral components of raising awareness and inspiring action on critical issues. Advocates and activists work to draw attention to environmental problems, influence policy decisions, and promote sustainable practices. Their efforts often include public demonstrations, petitions, lobbying government officials, and using media platforms to amplify their message. Through these actions, they mobilize communities and governments to take action to protect the environment (Gottfried, 2018).

One of the key strategies used by environmental activists is grassroots organizing. Grassroots movements are driven by community members who are directly affected by environmental degradation. These local efforts are often more powerful than top-down approaches, as they focus on issues that directly impact the well-being of individuals and their communities. Activists leverage social networks, local media, and community meetings to engage people and build momentum for change (Gottfried, 2018).

The success of environmental activism depends on the ability to educate the public about the urgency of the issue at hand. Advocacy groups often use scientific research, expert testimony, and compelling narratives to persuade individuals and decision-makers to act. By raising awareness about the consequences of environmental harm, activists can build support for policy reforms that protect natural resources, reduce pollution, and address climate change (Smith, 2020).

Integrating Sustainability into Daily Life

Educating individuals about the importance of sustainability is essential in encouraging them to adopt sustainable practices in their daily lives. Simple actions, such as reducing waste, conserving water, recycling, and using energy-efficient appliances, can make a significant difference in reducing environmental impact. Education programs that promote sustainable lifestyles help people understand that their choices, no matter how small, contribute to the overall health of the planet (Miller & Lane, 2019).

Sustainability education often focuses on teaching individuals how to reduce their carbon footprint, supporting local and organic food systems, and adopting green energy solutions. It empowers individuals to take ownership of their environmental impact and motivates them to make choices that benefit both the planet and their communities. For example, community-based initiatives that educate people on sustainable gardening, zero-waste lifestyles, and renewable energy options have proven successful in reducing environmental harm at the local level (Satterfield et al., 2020).

Furthermore, sustainable living education promotes the concept of intergenerational equity. By educating people about the long-term implications of environmental degradation, individuals are encouraged to think about the impact of their actions on future generations. This fosters a collective sense of responsibility and ensures that sustainability practices are passed down to future generations, creating a ripple effect of positive change across societies (Miller & Lane, 2019).

The Role of Governments in Education and Awareness

Governments play a critical role in shaping the educational landscape, especially in areas related to environmental issues and sustainability. Through policies and initiatives, governments can ensure that education systems include topics such as climate change, conservation, and sustainable development. Government-led educational programs often provide the structure and resources needed to raise awareness on a national and global scale (UNESCO, 2018).

For example, many countries have integrated environmental education into their national curricula, ensuring that future generations understand the importance of protecting the environment. Governments also fund public awareness campaigns that highlight critical issues like deforestation, overfishing, and plastic pollution. These campaigns help to inform the general public and encourage people to adopt more sustainable behaviors in their everyday lives.

In addition, governments can incentivize schools, businesses, and communities to participate in sustainability initiatives. For instance, offering tax breaks for organizations that invest in renewable energy or provide environmental education resources can encourage more widespread engagement with sustainability goals. Furthermore, by investing in research and innovation, governments can support the development of new technologies and solutions that can help address global challenges such as climate change (UNESCO, 2018).

The Impact of Environmental Education on Policy

Environmental education does not only impact individuals but can also influence policy decisions and government action. By educating the public and decision-makers about environmental issues, education can drive policy changes that prioritize sustainability and protect natural resources. Well-informed citizens are more likely to demand environmental policies that address the challenges of climate change, pollution, and biodiversity loss (Wallerstein, 2019).

In many cases, grassroots movements and public awareness campaigns have led to significant policy shifts. For instance, public pressure on governments to address plastic waste has resulted in the implementation of plastic bag bans and the promotion of plastic alternatives in several countries. Similarly, the growing awareness of climate change has influenced national governments to adopt renewable energy policies and pledge to reduce greenhouse gas emissions (Smith, 2020).

Environmental education programs can also provide decision-makers with the scientific knowledge they need to make informed policy choices. By incorporating research, case studies, and expert testimony into educational campaigns, advocates can build the case for stronger environmental regulations and more ambitious climate goals. When policy and education go hand in hand, the result is a more sustainable and equitable future (Wallerstein, 2019).

The Power of Youth in Spreading Awareness

Young people have always been at the forefront of social and environmental movements. In recent years, youth-led movements such as Fridays for Future and the School Strike for Climate have gained global attention and mobilized millions of young people around the world to demand action on climate change. These movements show the power of education and awareness in inspiring young people to become passionate advocates for environmental justice (Thunberg, 2019).

Young people often bring fresh perspectives and innovative ideas to the table, which can challenge traditional thinking and lead to new solutions. Through social media and online platforms, youth activists can spread their message to a global audience, creating an international network of supporters. This demonstrates the crucial role of education in empowering young people to raise awareness and become leaders in addressing environmental and social issues (Thunberg, 2019).

Moreover, youth movements often emphasize the need for intergenerational action, highlighting that the future of the planet is at stake and that young people are not willing to wait for future generations to address the problems they will inherit. By educating and organizing young people, societies can create a generation of environmentally conscious citizens who will continue to advocate for sustainability and social change for years to come (Thunberg, 2019).

Media and Its Role in Environmental Awareness

The media plays an essential role in spreading environmental awareness and educating the public about pressing issues such as climate change, pollution, and biodiversity loss. Traditional media outlets, such as television, newspapers, and radio, as well as digital platforms like blogs, podcasts, and social media, all contribute to the dissemination of critical information (Anderson, 2019). Through documentaries, news coverage, and investigative journalism, media outlets bring attention to environmental problems that might otherwise be overlooked.

One of the significant advantages of media in environmental awareness is its ability to reach large, diverse audiences. Stories about environmental issues can be broadcast to millions, which raises public consciousness and motivates action. For instance, documentaries such as *An Inconvenient Truth* and *The 11th Hour* have sparked widespread debates on climate change and sustainability. These films have not only educated viewers but also influenced policy discussions and led to changes in environmental laws (Smith, 2020).

Social media platforms also play a crucial role in mobilizing people for environmental causes. Activists use Twitter, Instagram, and Facebook to organize events, spread information, and advocate for policy change. Online petitions, viral campaigns, and digital activism have become powerful tools for raising awareness and influencing public opinion on environmental issues. With more people turning to digital spaces for news and information, the role of media in environmental education continues to grow, influencing both individuals and larger societal changes.

Local Communities as Catalysts for Change

Local communities are vital players in the movement toward environmental sustainability. Grassroots initiatives that focus on local environmental issues often create a direct, tangible impact on the surrounding area. Whether it is through neighborhood clean-ups, community gardens, or local recycling programs, these efforts can help reduce waste, conserve resources, and promote biodiversity at the local level (Satterfield et al., 2020).

Local communities are often in the best position to address environmental issues that directly affect their daily lives. They have the knowledge and experience necessary to develop innovative solutions that cater to the unique environmental challenges they face. For instance, in areas where water scarcity is a problem, local communities might implement rainwater harvesting systems or adopt water-efficient farming techniques. These small-scale, community-driven solutions can serve as models for other communities and contribute to a larger shift toward sustainable practices (Satterfield et al., 2020).

In addition to local efforts, communities also play a key role in advocating for policy change. When citizens mobilize at the local level, they can apply pressure to governments and corporations to adopt environmentally friendly policies and practices. The power of grassroots activism has been demonstrated through movements that have successfully opposed harmful industrial practices and lobbied for better environmental laws (Gottfried, 2018). By empowering local communities, societies can create a collective force that drives environmental change from the ground up.

Corporate Responsibility in Spreading Environmental Awareness

Businesses and corporations are increasingly recognizing their role in spreading environmental awareness and contributing to sustainable practices. Many companies have adopted sustainability as part of their corporate social responsibility (CSR) strategy, aligning their business practices with environmental goals. These efforts can include reducing carbon emissions, improving energy efficiency, and implementing waste reduction programs (Smith & Parker, 2020). By incorporating sustainability into their operations, businesses not only benefit the environment but also contribute to raising awareness about the importance of environmental protection.

Large corporations often have the resources and global reach to spread awareness on a massive scale. For example, multinational companies like Unilever and Tesla use their platforms to promote sustainable products and raise awareness about climate change. They invest in advertising campaigns and corporate partnerships to spread messages about the need for sustainable consumption and the importance of addressing environmental issues. In doing so, they encourage their customers, employees, and other stakeholders to consider the environmental impact of their choices (Smith & Parker, 2020).

Moreover, businesses can collaborate with nonprofits, educational institutions, and governments to develop and implement sustainability initiatives. Corporate partnerships with environmental organizations can create powerful synergies that amplify environmental education and awareness campaigns. By working together, businesses and environmental advocates can reach wider audiences and drive more substantial change.

Environmental Policies and Government Action

Governments have a pivotal role in raising environmental awareness through the creation and implementation of policies that protect the environment and encourage sustainable practices. By setting regulations, incentivizing sustainable practices, and promoting environmental education, governments can effectively drive public awareness and action (Wallerstein, 2019). Environmental policies often form the foundation for larger societal changes, influencing everything from industrial practices to personal behavior.

Policies such as carbon pricing, renewable energy incentives, and pollution control regulations are examples of how governments can directly influence the behavior of businesses and individuals. Furthermore, many governments use their legislative power to set long-term environmental goals, such as reducing greenhouse gas emissions and achieving net-zero carbon emissions by a certain date (Smith, 2020). These policies send a clear message to the public and businesses about the importance of sustainability and the need to reduce environmental harm.

Governments can also partner with international organizations, such as the United Nations, to raise awareness about global environmental challenges. Through these collaborations, countries can share best practices and work together to address transnational issues such as climate change and biodiversity loss. International cooperation on environmental issues enhances the effectiveness of policies and fosters global solidarity in tackling climate change (Wallerstein, 2019).

Sustainable Practices in the Corporate World

Sustainability has become a key focus in the corporate world, with businesses across various industries adopting practices that minimize environmental impact. This shift toward sustainability is partly driven by consumer demand for greener products and services, as well as increasing regulatory pressures to address environmental issues. Companies that adopt sustainable practices not only reduce their environmental footprint but also contribute to the growing global movement for environmental awareness (Anderson, 2019).

Many businesses have implemented initiatives to reduce waste, lower energy consumption, and adopt renewable energy sources. For example, companies like Google and Microsoft have committed to running their operations entirely on renewable energy, while others focus on reducing water usage or implementing circular economy models. By prioritizing sustainability, these companies help raise awareness about the importance of reducing environmental impact and demonstrate the potential for businesses to drive positive environmental change.

Corporate sustainability efforts can also have a ripple effect, encouraging other companies to adopt similar practices and influencing the behavior of consumers. Through marketing campaigns, partnerships with environmental organizations, and sustainability reports, companies share their sustainability efforts with the public, raising awareness about the need for greener business practices. As more businesses lead by example, sustainable practices become more ingrained in industries worldwide.

Education Systems and Environmental Curriculum

Education systems around the world are beginning to prioritize environmental literacy as a core component of their curricula. This shift towards incorporating sustainability education aims to equip students with the knowledge and skills needed to address global environmental challenges, including climate change, biodiversity loss, and resource depletion (UNESCO, 2020). By teaching students about environmental issues and encouraging critical thinking about sustainability, education systems can foster a generation of informed citizens who are ready to tackle environmental challenges.

In many countries, schools are introducing subjects that focus specifically on environmental issues. Topics such as climate science, renewable energy, sustainable agriculture, and environmental ethics are becoming common in primary and secondary education. In addition, many universities now offer programs in environmental studies, providing students with a deeper understanding of environmental science and policy (Miller & Lane, 2019).

The integration of environmental education into school curricula also offers the opportunity to promote sustainable behavior from an early age. By teaching children about the importance of conservation, recycling, and reducing waste, schools can instill lifelong habits that help protect the planet. Moreover, experiential learning opportunities, such as school gardens or sustainability projects, allow students to apply what they have learned in real-world settings and make a tangible impact in their communities (UNESCO, 2020).

The Future of Environmental Education

The future of environmental education looks promising, as both global awareness and the urgency of environmental issues continue to grow. Education systems, governments, and communities are increasingly recognizing the importance of sustainability and environmental literacy in preparing future generations to face the environmental challenges ahead. This shift towards environmental education is not just about knowledge transfer but also about empowering individuals to take action and create lasting change.

In the future, environmental education will likely become even more integrated into all aspects of learning. From STEM (Science, Technology, Engineering, and Mathematics) programs to art and social studies, sustainability principles will be embedded across disciplines. This interdisciplinary approach will help students understand the interconnectedness of environmental, social, and economic issues and equip them with the tools to solve complex global problems (Miller & Lane, 2019).

Moreover, advancements in technology will allow for more innovative and engaging methods of environmental education. Virtual reality (VR) and augmented reality (AR) tools, for instance, can bring immersive experiences that help learners better understand environmental issues, such as climate change impacts or biodiversity conservation efforts. As technology continues to evolve, educational systems will increasingly use these tools to create interactive learning environments that inspire action and promote sustainable practices.

Ultimately, the future of environmental education is bright, and it will play a crucial role in creating a more sustainable, just, and resilient world for future generations.

Chapter 17: Personal Choices: Everyday actions for change

Introduction to Personal Choices and Environmental Change

Personal choices have a profound impact on the environment, as everyday actions collectively shape global trends in resource consumption, waste production, and energy use. From the food we eat to the transportation we use, every decision influences the environment in both direct and indirect ways (Smith & Clark, 2020). While individual actions might seem small in isolation, when multiplied by billions of people, they can contribute to significant environmental changes. Thus, understanding the power of personal choices is crucial for fostering a more sustainable future.

The principle of sustainable living emphasizes the importance of making decisions that meet the needs of the present without compromising the ability of future generations to meet their own needs (United Nations, 1987). Personal choices that align with sustainability goals include actions such as reducing waste, conserving water, using renewable energy, and supporting eco-friendly products. These choices are not only beneficial for the environment but can also lead to improved personal well-being and cost savings.

Environmental experts have stressed that fostering awareness and encouraging positive behavioral changes are essential components of achieving global sustainability goals. By understanding the impact of everyday actions, individuals can make informed decisions that reduce their environmental footprint. Simple actions, such as reducing plastic use, adopting plant-based diets, or choosing public transportation, can make a substantial difference when practiced by many individuals. The cumulative impact of these personal choices can contribute to the reduction of greenhouse gas emissions, conservation of natural resources, and the preservation of biodiversity (Johnson et al., 2020).

Reducing Carbon Footprint: Simple Steps for Individuals

One of the most significant personal choices individuals can make to reduce their environmental impact is minimizing their carbon footprint. The carbon footprint refers to the total amount of greenhouse gases, particularly carbon dioxide (CO2), emitted as a result of human activities. By adopting energy-efficient practices, individuals can lower their contribution to global warming and reduce their reliance on fossil fuels.

A key step in reducing one's carbon footprint is to decrease energy consumption at home. This can be achieved by using energy-efficient appliances, such as LED light bulbs, and by ensuring that heating and cooling systems are optimized. Additionally, unplugging electronics when not in use and using a programmable thermostat can help conserve energy. According to the U.S. Environmental Protection Agency (EPA), making simple changes in the home, such as turning off lights when leaving a room or installing insulation, can significantly reduce energy consumption and cut carbon emissions (EPA, 2019).

Transportation is another area where individuals can make a noticeable impact. Cars are a major source of carbon emissions, particularly those that run on gasoline. Choosing public transportation, carpooling, or cycling instead of driving alone can drastically reduce carbon emissions. Additionally, switching to electric vehicles (EVs) or hybrid cars, which produce fewer emissions than traditional cars, is becoming increasingly popular as technology improves (Johnson et al., 2020). By making environmentally conscious transportation choices, individuals contribute to a reduction in the demand for fossil fuels and help lower overall greenhouse gas emissions.

Water Conservation: The Impact of Everyday Choices

Water is one of the planet's most precious resources, yet it is often taken for granted in daily life. Personal choices can play a significant role in conserving water, reducing waste, and ensuring that this vital resource is available for future generations. According to the United Nations, water scarcity affects more than 2 billion people worldwide, making water conservation efforts even more important (United Nations, 2020). By making small adjustments to everyday habits, individuals can help mitigate this global issue.

One of the simplest ways to conserve water is to reduce consumption in daily activities. For example, turning off the faucet while brushing teeth or washing dishes can save gallons of water each day. Similarly, fixing leaky pipes and faucets can prevent water wastage and lower utility bills (EPA, 2019). Installing water-efficient appliances, such as low-flow showerheads, faucets, and toilets, can also reduce water usage significantly. These small changes, when implemented in households across the world, can result in a substantial reduction in water consumption.

Moreover, individuals can choose to use water-saving gardening practices, such as watering plants during early morning hours to reduce evaporation, or installing rain barrels to collect and reuse rainwater for outdoor irrigation. These choices can help conserve water and ensure that it is used more efficiently. In addition to these practices, promoting water conservation in communities and supporting policies aimed at protecting freshwater resources can further support global efforts to address water scarcity (Johnson et al., 2020).

Reducing Waste: The Power of Recycling

Waste management is a pressing issue that individuals can address through everyday choices. By reducing waste and recycling, people can significantly reduce the strain on landfills, conserve natural resources, and lower environmental pollution. Recycling not only prevents valuable materials from being thrown away but also reduces the need for raw materials, thus conserving energy and reducing environmental harm caused by extraction and processing.

One of the most straightforward ways to contribute to waste reduction is by practicing recycling. Many everyday items, such as paper, cardboard, plastic, and glass, can be recycled instead of being sent to the landfill. By separating recyclables from non-recyclables and ensuring that they are disposed of correctly, individuals can help divert waste from landfills and reduce pollution. In addition, composting organic waste, such as food scraps and yard waste, reduces the need for landfill space and provides natural fertilizer for gardens and crops (EPA, 2019).

Another important action is reducing consumption and waste before it even begins. By choosing products with minimal packaging, purchasing in bulk, and opting for reusable items such as cloth bags, water bottles, and food containers, individuals can decrease the amount of waste they generate in the first place. Supporting companies that use sustainable packaging and promote circular economy models also helps foster environmentally responsible business practices.

Sustainable Eating Habits: The Role of Diet in Environmental Change

One of the most impactful personal choices individuals can make is in their dietary habits. Food production, especially the farming of livestock, has significant environmental implications, from the use of water and land to greenhouse gas emissions. Adopting more sustainable eating habits can help reduce one's carbon footprint and contribute to a healthier planet (Smith & Clark, 2020).

A plant-based diet, for example, has a lower environmental impact compared to a meat-heavy diet. The production of plant-based foods, such as fruits, vegetables, grains, and legumes, requires fewer resources and produces fewer emissions compared to animal agriculture (Garnett, 2019). Reducing meat consumption, particularly beef and lamb, can significantly lower an individual's carbon footprint. Additionally, choosing locally grown and seasonal foods helps minimize the environmental impact of transportation and supports sustainable farming practices (Miller & Lane, 2019).

Food waste is another major concern that individuals can address through their choices. According to the Food and Agriculture Organization (FAO), approximately one-third of all food produced globally is wasted, contributing to unnecessary greenhouse gas emissions and resource depletion (FAO, 2020). By planning meals, storing food properly, and using leftovers, individuals can significantly reduce food waste and contribute to a more sustainable food system.

Lastly, supporting organic and regenerative farming practices promotes soil health, reduces pesticide use, and encourages biodiversity. By choosing organic produce and sustainably sourced products, individuals can help foster a more environmentally responsible agricultural system.

Supporting Sustainable Products: Ethical Consumerism

Ethical consumerism is an approach where individuals make purchasing decisions that reflect their values, particularly those related to environmental sustainability and social justice. By choosing products that are produced with minimal environmental impact and under fair labor conditions, consumers can support companies that align with their ethical standards (Johnson et al.,

2020). Ethical consumerism empowers individuals to influence market trends and encourage corporations to adopt more sustainable and responsible practices.

One area where consumer choices can make a substantial difference is in the selection of eco-friendly products. Items such as clothing, electronics, and household goods are often made from non-renewable resources and may be associated with harmful environmental practices. Opting for products made from recycled materials, or those that are designed for longevity and repair, reduces waste and encourages manufacturers to adopt sustainable production processes. Certifications such as Fair Trade and organic labels also help consumers identify products that meet higher environmental and social standards.

Moreover, individuals can support sustainable businesses by shopping locally and supporting companies that prioritize sustainability in their operations. By purchasing from local farmers' markets, fair trade shops, and small-scale artisans, consumers help reduce the environmental impact of shipping and transportation, while supporting businesses that invest in sustainable practices.

Ethical consumerism also extends to the choices made when discarding items. Instead of opting for disposable products, individuals can invest in reusable goods, reducing the demand for single-use plastics and minimizing landfill waste. Through these conscious purchasing decisions, individuals contribute to a more sustainable global economy (Johnson et al., 2020).

Energy Conservation: Saving Power in Daily Life

Energy conservation is a critical element in reducing environmental impact, as the production of energy, particularly from fossil fuels, contributes significantly to air pollution and climate change. Individuals can play a pivotal role in conserving energy by implementing simple changes to their daily habits and investing in energy-efficient technologies (Smith & Clark, 2020).

One of the most effective actions individuals can take to conserve energy is to reduce heating and cooling costs. Setting thermostats to a lower temperature in winter and a higher temperature in summer, or using fans and natural ventilation, can minimize the need for air conditioning and heating systems, which are major energy consumers. Additionally, investing in insulation for homes and using weather stripping to seal drafts can reduce the loss of heat in winter and keep homes cooler in the summer, leading to significant energy savings (EPA, 2019).

Another way to conserve energy is by using energy-efficient appliances. Appliances that carry the ENERGY STAR label use less energy to perform the same tasks, reducing both electricity consumption and environmental impact. Replacing old appliances with newer, energy-efficient models or choosing LED bulbs over incandescent lights can further reduce energy use.

On a larger scale, individuals can also advocate for the adoption of renewable energy sources in their communities. Supporting policies and initiatives that promote the use of solar, wind, and other renewable sources can lead to a reduction in reliance on fossil fuels and a cleaner, more sustainable energy future.

Embracing Public Transportation: Reducing Carbon Emissions

Public transportation plays a vital role in reducing the environmental impact of individual travel. As cities become increasingly urbanized, the demand for transportation solutions that minimize emissions and reduce traffic congestion grows. Opting for public transit instead of private vehicles can significantly lower the amount of greenhouse gases produced by transportation, as buses, trains, and trams are more energy-efficient on a per-person basis (Johnson et al., 2020).

Using public transportation reduces the number of private vehicles on the road, which in turn lowers traffic congestion and reduces air pollution. In many cities, public transit systems are powered by electricity, which can be generated from renewable sources such as wind or solar energy, further reducing emissions (Smith & Clark, 2020). Additionally, public transit options, including shared carpooling programs, decrease the need for multiple single-occupancy vehicles, which are less energy-efficient and contribute to higher levels of air pollution.

Individuals who regularly use public transportation not only help reduce their own carbon footprints but also support the development of sustainable urban infrastructure. By advocating for improved transit systems and increased investment in environmentally friendly transportation options, individuals can contribute to building more sustainable cities. Public transportation is a key strategy in reducing urban emissions and making cities more livable and environmentally friendly.

Furthermore, choosing to cycle or walk short distances instead of driving further reduces reliance on fossil fuels and contributes to personal health and well-being.

Reducing Plastic Usage: A Key to Environmental Protection

Plastic pollution has become one of the most pressing environmental issues of the 21st century. Single-use plastics, such as bags, bottles, and packaging, contribute to pollution in oceans, rivers, and landfills, where they can persist for hundreds of years. By making conscious choices to reduce plastic usage, individuals can significantly mitigate the harmful effects of plastic pollution and help protect ecosystems and wildlife (Garnett, 2019).

One of the most effective steps in reducing plastic waste is to avoid single-use plastics altogether. Instead of using disposable plastic bags, individuals can carry reusable cloth bags for shopping. Opting for reusable containers, such as stainless-steel water bottles or glass jars, instead of plastic bottles or cups, reduces the demand for plastic products (EPA, 2019). Many companies and businesses are now offering alternatives to plastic packaging, such as biodegradable or recyclable materials, allowing consumers to make more sustainable choices.

Furthermore, individuals can support businesses and organizations that are committed to reducing plastic waste. By choosing to purchase from companies that prioritize sustainability and use eco-friendly packaging, consumers can send a message to the market, encouraging the adoption of greener practices across industries (Johnson et al., 2020). Engaging in recycling and supporting policies aimed at reducing plastic production are additional ways individuals can take action against plastic pollution.

In conclusion, reducing plastic usage is an important step in minimizing environmental harm and fostering a more sustainable future for the planet.

Supporting Sustainable Fashion: Ethical Clothing Choices

The fashion industry has significant environmental and social impacts, from resource extraction to waste production. However, individual choices in purchasing and wearing clothing can help shift the industry toward more sustainable practices. Sustainable fashion emphasizes eco-friendly production methods, ethical labor practices, and minimal environmental impact (Miller & Lane, 2019). By making informed clothing choices, consumers can contribute to a cleaner and more ethical fashion industry.

One way to support sustainable fashion is by choosing clothing made from natural, organic, or recycled fibers. Fabrics such as organic cotton, hemp, and bamboo require fewer pesticides and chemicals to produce, reducing the environmental impact compared to conventional cotton or synthetic fibers. Additionally, buying from brands that prioritize ethical labor practices and fair wages for workers ensures that clothing production does not come at the expense of human rights.

Another important aspect of sustainable fashion is the longevity of clothing. Instead of frequently purchasing fast fashion items that are cheaply made and quickly discarded, individuals can invest in high-quality pieces that last longer and can be repaired if damaged. Choosing second-hand or vintage clothing also reduces the demand for new garments and extends the lifespan of existing clothing (Johnson et al., 2020).

Lastly, individuals can take care of their clothing by washing clothes in cold water, air-drying them, and avoiding excessive ironing. These simple practices can prolong the lifespan of garments and reduce energy consumption associated with clothing maintenance.

Water Conservation: Simple Actions for a Sustainable Future

Water is one of the most precious resources on the planet, yet it is often used wastefully in daily life. Freshwater scarcity is becoming an increasingly pressing issue, especially in regions facing droughts or over-extraction. However, individuals can play a significant role in conserving water by adopting simple habits that reduce consumption and waste (Smith & Clark, 2020).

One of the most effective actions individuals can take to conserve water is reducing water usage in daily tasks. This includes simple steps such as turning off the tap while brushing teeth or washing dishes, and using low-flow showerheads and faucets. Installing water-efficient appliances, such as washing machines and dishwashers that consume less water per cycle, can further help reduce water usage in households.

Another way to conserve water is by collecting rainwater for use in gardens and landscaping. Rainwater harvesting reduces the need to use potable water for irrigation and helps manage stormwater runoff, which can pollute rivers and lakes. Additionally, individuals can choose drought-resistant plants for their gardens, which require less water and can thrive in arid conditions (EPA, 2019).

Conserving water also extends to mindful consumption of bottled water. By choosing reusable water bottles and filtering tap water, individuals can reduce the demand for plastic bottles, which contribute to pollution and waste (Garnett, 2019).

Lastly, raising awareness about the importance of water conservation can help inspire others to take action. Collective efforts, both individual and community-driven, can significantly reduce water consumption and safeguard this vital resource for future generations.

Sustainable Gardening: Growing Your Own Food

Gardening is not only a rewarding and therapeutic activity, but it also contributes to a sustainable lifestyle by promoting local food production. Growing one's own food helps reduce dependence on industrial agriculture, decreases transportation emissions, and encourages the consumption of organic, pesticide-free produce (Miller & Lane, 2019).

Even those with limited space can practice sustainable gardening. Urban dwellers can grow herbs, vegetables, and small fruits in containers on balconies, windowsills, or rooftops. Container gardening allows individuals to produce food in small, manageable spaces while reducing the environmental impact of food transportation. Additionally, urban gardening fosters a greater connection to the food source, helping individuals understand where their food comes from and appreciate its value.

For those with more outdoor space, raised bed gardening and permaculture techniques allow for efficient use of land and resources. Permaculture emphasizes the use of natural ecosystems and sustainable practices to create self-sufficient gardens that require minimal input from external resources (Johnson et al., 2020).

Composting is another way to integrate sustainability into gardening. By turning organic waste into compost, individuals can enrich their soil naturally, reducing the need for chemical fertilizers. Composting also reduces waste sent to landfills, where it would otherwise produce harmful greenhouse gases (EPA, 2019).

In conclusion, sustainable gardening not only supports the environment but also empowers individuals to take control of their food production and reduce their ecological footprint.

Reducing Meat Consumption: A Simple but Powerful Change

Reducing meat consumption is one of the most impactful changes individuals can make to reduce their environmental footprint. Meat production, particularly from livestock such as cows and sheep, requires substantial amounts of land, water, and energy. Additionally, the livestock industry is a significant contributor to greenhouse gas emissions, particularly methane, which is much more potent than carbon dioxide (Garnett, 2019).

Individuals can reduce their meat consumption by adopting plant-based alternatives and incorporating more vegetables, grains, and legumes into their diets. For example, switching from beef to plant-based proteins like beans, lentils, or tofu reduces the demand for land and water resources used in livestock farming. Plant-based foods also generally require fewer fossil fuels for transportation and processing (Smith & Clark, 2020).

Even those who are not ready to completely eliminate meat from their diets can adopt strategies to reduce its environmental impact. For example, consuming smaller portions of meat, choosing sustainably sourced or grass-fed meats, and reducing food waste by utilizing leftovers can help mitigate the environmental effects of meat consumption (Miller & Lane, 2019).

Additionally, adopting a local and seasonal diet that emphasizes foods produced closer to home can further reduce the carbon footprint associated with meat and other food products. By making more sustainable food choices, individuals can contribute to a more environmentally responsible food system.

Embracing Minimalism: Reducing Consumerism

Adopting a minimalist lifestyle is an effective way to reduce personal consumption and minimize one's environmental impact. Minimalism emphasizes the value of living with less—focusing on quality over quantity and making intentional choices about the items one purchases. This mindset can have a profound effect on both the individual and the planet.

By embracing minimalism, individuals can reduce their demand for goods that are often produced unsustainably and contribute to environmental degradation. For instance, purchasing fewer clothing items reduces the need for fast fashion, which is a major contributor to textile waste and pollution. Instead, individuals can focus on investing in durable, high-quality pieces that will last longer and can be repaired when necessary (Smith & Clark, 2020).

Minimalism also encourages the reduction of electronic waste. Instead of frequently upgrading gadgets and electronics, minimalists tend to repair or repurpose their existing items, helping to extend their life cycle and reduce landfill waste (Garnett, 2019).

Furthermore, minimalism promotes sustainable living by fostering a greater appreciation for the resources already in one's possession. This shift in mindset can lead individuals to rethink unnecessary purchases and shift their focus toward experiences or long-lasting, meaningful possessions. It also encourages the use of shared resources and reducing one's overall carbon footprint by avoiding the overconsumption of products that are often wasteful.

In conclusion, by embracing minimalism, individuals can help reduce waste, conserve resources, and contribute to a more sustainable world.

Supporting Renewable Energy: Choosing Green Power

As the world faces the growing threat of climate change, the shift to renewable energy has become a crucial component of a sustainable future. Individuals can contribute to this transition by supporting renewable energy sources, such as solar, wind, and hydroelectric power, and making choices that reduce their reliance on fossil fuels (Johnson et al., 2020).

One of the most direct ways individuals can support renewable energy is by choosing green power options for their homes. Many utility providers now offer renewable energy plans, allowing consumers to opt for electricity generated from solar or wind sources. By making this choice, individuals contribute to the demand for cleaner energy and help accelerate the transition away from coal and natural gas.

In addition to switching to renewable energy providers, individuals can reduce their overall energy consumption by adopting energy-efficient practices. Using LED light bulbs, unplugging electronics when not in use, and ensuring homes are well-insulated can help reduce the amount of energy required to power a household (EPA, 2019). These actions not only reduce the demand for energy but also lower overall carbon emissions.

Furthermore, individuals can invest in renewable energy technologies such as solar panels or wind turbines for their homes. While initial costs may be high, the long-term benefits in terms of energy savings and reducing carbon footprints make this an increasingly popular option for homeowners (Garnett, 2019).

By making informed decisions about energy consumption and supporting renewable energy sources, individuals play a vital role in addressing climate change.

Engaging in Eco-Friendly Travel: Sustainable Tourism Choices

Travel and tourism contribute significantly to global carbon emissions due to the transportation sector, especially air travel. However, individuals can make more eco-friendly travel choices by adopting sustainable tourism practices that reduce their carbon footprint while still enjoying the benefits of exploration (Smith & Clark, 2020).

One of the most effective ways to reduce the environmental impact of travel is by choosing sustainable transportation options. For example, traveling by train, bus, or carpooling rather than flying reduces the carbon emissions associated with long-distance travel. Train journeys, especially in regions with extensive rail networks, tend to be more energy-efficient than air travel and are a great way to reduce one's environmental footprint.

Additionally, individuals can opt for eco-friendly accommodations, such as hotels or guesthouses that follow sustainable practices, including energy efficiency, waste reduction, and water conservation. Supporting local businesses that prioritize sustainability, including eco-tourism operators, helps ensure that tourism does not negatively impact local environments and communities (Garnett, 2019).

Sustainable travel also involves being mindful of one's impact on natural and cultural resources. Travelers should aim to minimize waste by avoiding single-use plastics and respecting local customs and ecosystems. Choosing to support wildlife conservation efforts and community-based tourism initiatives ensures that tourism benefits both the environment and local populations (Johnson et al., 2020).

Ultimately, by making informed and sustainable choices, individuals can reduce their environmental impact while still enjoying the rewards of travel and cultural exchange.

Sustainable Fashion: Choosing Eco-Friendly Clothing

The fashion industry is one of the largest contributors to global environmental degradation. From the excessive use of water and chemicals in textile production to the waste generated by fast fashion, the environmental cost of clothing is often overlooked (Garnett, 2019). However, individuals can make a significant impact by adopting more sustainable fashion practices.

One of the most effective ways to contribute to a more sustainable fashion industry is by purchasing clothing made from eco-friendly materials. Fabrics such as organic cotton, hemp, and bamboo require fewer pesticides and water resources to produce compared to conventional cotton or synthetic fibers. Moreover, clothing made from recycled materials, such as recycled polyester, can help reduce the demand for virgin resources.

Another important consideration is buying fewer, but higher-quality, items. Fast fashion encourages the purchase of cheap, disposable clothing that quickly wears out, resulting in more textile waste. By investing in durable, timeless pieces that can be worn for years, individuals can reduce the environmental burden associated with frequent clothing purchases (Miller & Lane, 2019).

Additionally, supporting brands that prioritize sustainability in their production processes and supply chains can make a substantial difference. Many clothing brands are now adopting ethical production practices, using sustainable materials, and minimizing waste (Smith & Clark, 2020). Consumers who choose to support these brands send a clear message that sustainability matters and encourage the industry to shift towards more responsible practices.

By being mindful of clothing choices and focusing on sustainability, individuals can help reduce the environmental impact of the fashion industry.

Sustainable Investing: Putting Your Money Where It Matters

Investing is an often-overlooked aspect of sustainability, but individuals have the power to make a positive impact on the environment through their financial choices. Sustainable investing, also known as impact investing or socially responsible investing (SRI), involves choosing investments that prioritize environmental, social, and governance (ESG) criteria (Johnson et al., 2020). By supporting companies that are committed to sustainability, investors can promote positive change on a larger scale.

One way individuals can engage in sustainable investing is by focusing on companies that actively work toward environmental protection, renewable energy adoption, or waste reduction. Many companies today are committing to net-zero carbon emissions, making efforts to reduce their environmental footprints, and offering environmentally friendly products and services. Investing in such companies encourages other businesses to adopt similar practices.

Additionally, individuals can consider investing in green bonds or funds, which are specifically designed to finance environmentally sustainable projects such as clean energy, water conservation, and waste management (Miller & Lane, 2019). These investment vehicles not only provide financial returns but also support initiatives that align with personal values and contribute to a more sustainable future.

Sustainable investing also involves avoiding investments in industries that harm the environment, such as fossil fuels, tobacco, and deforestation-related activities. By divesting from harmful sectors and focusing on businesses that prioritize sustainability, individuals can take a more active role in shaping the economy in a way that supports the planet's long-term health.

Reducing Food Waste: A Simple Step for a Greener World

Food waste is a major environmental issue that contributes to unnecessary resource depletion and increases greenhouse gas emissions. According to the Food and Agriculture Organization (FAO), approximately one-third of all food produced globally is wasted, which equates to significant energy, water, and land resources being squandered (FAO, 2019). However, individuals can take simple steps to reduce food waste and promote a more sustainable food system.

One way to reduce food waste is by planning meals and purchasing only the necessary ingredients. Many people overbuy food, which often leads to waste when it goes uneaten. By making shopping lists, checking inventory before purchasing, and buying in smaller quantities, individuals can prevent excess food from going to waste (Garnett, 2019).

Another method for reducing food waste is by properly storing food to extend its shelf life. Techniques such as freezing perishable items, storing fruits and vegetables in the right conditions, and using leftovers creatively can all help reduce food waste.

Composting is another important practice for individuals who want to reduce food waste. Rather than sending food scraps to landfills, composting allows organic waste to break down into nutrient-rich soil that can be used in gardens and landscaping. This practice not only reduces waste but also enhances soil health and reduces the need for chemical fertilizers (EPA, 2019).

By making small, mindful changes in food purchasing, storage, and disposal practices, individuals can significantly reduce food waste and contribute to a more sustainable food system.

Supporting Local and Organic Food Systems

Supporting local and organic food systems is another impactful way for individuals to make more sustainable choices. By purchasing food from local farmers, individuals can reduce the environmental costs associated with long-distance transportation, packaging, and refrigeration. In addition, local farmers often employ more sustainable farming practices that are less reliant on chemicals and synthetic fertilizers, which helps preserve the health of the land and surrounding ecosystems (Barker, 2021).

One way to support local food systems is by shopping at farmers' markets, where consumers can buy fresh produce directly from the growers. This direct relationship between producers and consumers reduces the need for middlemen and allows consumers to make informed choices about how their food is grown and harvested. Additionally, farmers' markets often feature seasonal produce that is fresher and requires fewer resources to grow (Smith & Clark, 2020).

Choosing organic foods is another significant way to contribute to sustainability. Organic farming avoids the use of synthetic pesticides and fertilizers, relying instead on natural methods to improve soil fertility and manage pests. This reduces pollution and the harmful impact of chemicals on the environment. Organic farming also typically involves crop rotation, which enhances soil health and biodiversity (Miller & Lane, 2019).

Furthermore, by reducing demand for mass-produced, industrially farmed food, individuals can support more ethical practices that prioritize animal welfare and ecological sustainability. Supporting local and organic food systems not only helps individuals live more sustainably but also promotes the broader adoption of sustainable agriculture practices.

Chapter 18: Success Stories: Learning from examples

Introduction to Success Stories in Sustainability

Success stories in sustainability provide valuable lessons on how innovative solutions can address environmental challenges. Around the world, governments, businesses, and communities have implemented strategies to reduce pollution, promote renewable energy, and protect natural ecosystems. By studying these examples, we can identify best practices and adapt them to different regions and contexts (Smith, 2021).

One reason success stories are crucial is that they demonstrate the feasibility of sustainability initiatives. Many environmental solutions face skepticism due to concerns about economic costs or technological limitations. However, examples like **Costa Rica's renewable energy program** and **San Francisco's Zero Waste initiative** prove that sustainability efforts can be both practical and beneficial (González, 2020).

Additionally, learning from real-world cases helps policymakers and organizations avoid common pitfalls. For instance, the Netherlands' flood management system provides insights into how coastal regions can prepare for rising sea levels. Meanwhile, Norway's electric vehicle incentives illustrate how government policies can accelerate the transition to cleaner transportation (Jansen & Petersen, 2019).

This chapter will explore multiple success stories from various sectors, including energy, conservation, waste management, and policy. Each case will highlight key takeaways, emphasizing the role of innovation, collaboration, and long-term planning in achieving sustainability goals.

Costa Rica's Renewable Energy Success

Costa Rica is a global leader in renewable energy, demonstrating that a country can power itself almost entirely with clean energy sources. In recent years, Costa Rica has consistently generated over **99% of its electricity from renewable sources**, including hydropower, wind, solar, and geothermal energy (Jiménez, 2021).

One of the key reasons for Costa Rica's success is its investment in **hydroelectric power**, which accounts for nearly **70% of the country's electricity** (World Bank, 2020). The country's abundant rivers and rainfall have enabled the development of large-scale hydroelectric dams, providing a stable and renewable energy supply. Additionally, Costa Rica has expanded **wind and geothermal energy projects**, further reducing reliance on fossil fuels.

Government policies have played a crucial role in this transition. The Costa Rican government has implemented **strong environmental regulations**, financial incentives for renewable energy projects, and a long-term commitment to sustainability (González, 2020). By prioritizing green energy, the country has achieved significant economic and environmental benefits, including **reduced carbon emissions and energy independence**.

Costa Rica's success offers valuable lessons for other nations seeking to transition to renewable energy. The case demonstrates the importance of **strong political commitment, strategic investments, and public support** in achieving sustainable energy goals. As global concerns over climate change grow, Costa Rica serves as a model for how clean energy can drive national progress.

San Francisco's Zero Waste Initiative

San Francisco has emerged as a global leader in waste management, implementing an ambitious **Zero Waste initiative** that aims to **divert 100% of waste from landfills** by recycling and composting (Zaman & Lehmann, 2019). As of 2020, the city has successfully diverted over **80% of its waste**, making it one of the most sustainable urban centers in the world (San Francisco Department of the Environment, 2021).

A key factor behind San Francisco's success is its **mandatory composting and recycling laws**. In 2009, the city passed the **Mandatory Recycling and Composting Ordinance**, requiring all residents and businesses to separate recyclables and organic waste from landfill trash (Smith, 2021). This policy has significantly reduced waste and increased compost production.

Public education campaigns have also played a crucial role. The city has invested in **community outreach programs**, educating residents about the importance of waste reduction and providing guidance on proper sorting methods (Zaman & Lehmann, 2019). These efforts have increased public participation and improved recycling rates.

Additionally, San Francisco has **banned plastic bags, Styrofoam containers, and single-use plastics**, encouraging businesses and consumers to adopt reusable alternatives (San Francisco Department of the Environment, 2021). This has further reduced landfill waste and plastic pollution.

San Francisco's Zero Waste initiative demonstrates that **strong policies, community involvement, and strict regulations** can lead to significant environmental improvements. Other cities looking to improve waste management can learn from San Francisco's proactive approach.

The Netherlands' Flood Management Innovations

The Netherlands has developed one of the most advanced flood management systems in the world, demonstrating how **engineering and environmental planning** can protect communities from climate change-related disasters. With **over a quarter of its land below sea level**, the Netherlands has historically been vulnerable to flooding (Hooijer et al., 2020). However, through **innovative water**

management projects, the country has successfully minimized flood risks.

One of the most notable achievements is the **Delta Works**, a series of dams, sluices, and storm surge barriers designed to control water levels and prevent flooding (Deltacommissie, 2019). The **Maeslantkering storm surge barrier**, for example, automatically closes when sea levels rise too high, protecting major cities like Rotterdam.

In addition to physical barriers, the Netherlands has implemented a **"Room for the River"** program, which allows rivers to naturally overflow into designated areas, reducing pressure on flood defenses (Hooijer et al., 2020). This approach balances **engineering solutions with ecological preservation**, ensuring sustainable flood control.

Government policies and public awareness campaigns have further strengthened flood resilience. The Dutch government continually invests in **climate adaptation strategies**, and local communities actively participate in **water management planning** (Deltacommissie, 2019).

The Netherlands' flood management success provides valuable lessons for coastal regions worldwide. By integrating **technology, nature-based solutions, and public participation**, countries facing rising sea levels can develop sustainable strategies to protect their populations.

Norway's Electric Vehicle Revolution

Norway has emerged as a global leader in **electric vehicle (EV) adoption**, demonstrating how government policies and financial incentives can accelerate the transition to cleaner transportation. As of 2022, **nearly 80% of all new cars sold in Norway are electric**, making it the country with the highest EV market share worldwide (International Energy Agency [IEA], 2022).

A key factor behind Norway's EV success is its **comprehensive incentive program**. The Norwegian government has **eliminated purchase taxes and VAT on EVs**, making them more affordable than gasoline-powered cars (Jansen & Petersen, 2021). Additionally, EV owners benefit from perks such as **free public charging, reduced toll fees, and access to bus lanes**, further encouraging adoption.

Infrastructure investment has also played a crucial role. Norway has established an **extensive network of fast-charging stations**, allowing EV owners to travel long distances without range anxiety (IEA, 2022). The government continues to expand charging infrastructure in both urban and rural areas.

Public awareness and social acceptance have strengthened the transition. Norwegian consumers have embraced EVs due to their **cost savings, environmental benefits, and government support** (Jansen & Petersen, 2021). Furthermore, major automakers have prioritized Norway as a test market for new EV models, ensuring a diverse selection for consumers.

Norway's EV revolution serves as a **model for other nations** seeking to reduce carbon emissions in the transportation sector. By combining **financial incentives, infrastructure investment, and strong public support**, Norway has demonstrated that large-scale EV adoption is possible.

Rwanda's Reforestation Success

Rwanda has demonstrated how **reforestation efforts** can restore ecosystems and support economic development. After years of deforestation due to agriculture and population growth, the government launched an ambitious **forest restoration program**, increasing national forest cover from **17% in 1990 to over 30% by 2020** (Mukuralinda et al., 2021).

A key factor in Rwanda's success is its **Forest Landscape Restoration (FLR) initiative**, which aligns with the **Bonn**

Challenge, a global effort to restore **350 million hectares of degraded land** by 2030 (Global Landscapes Forum, 2020). This program integrates **tree planting with sustainable land use**, benefiting both biodiversity and local communities.

Government policies have played a vital role. Rwanda implemented **strict anti-deforestation laws**, requiring tree planting on farmlands and creating protected areas (Mukuralinda et al., 2021). Additionally, Rwanda promotes **agroforestry**, where trees and crops grow together, improving soil health and increasing farmers' resilience to climate change.

Community involvement has been critical. Rwanda's **Umuganda** (monthly community service day) encourages citizens to participate in tree planting, fostering a culture of **environmental stewardship** (Global Landscapes Forum, 2020). This approach ensures that reforestation efforts are locally supported and sustained.

Rwanda's success proves that **strong policies, community participation, and sustainable land management** can reverse deforestation and restore ecosystems. Other nations struggling with land degradation can learn from Rwanda's holistic approach to reforestation.

Germany's Renewable Energy Transition

Germany has emerged as a global leader in **renewable energy**, pioneering a large-scale transition away from fossil fuels through its **Energiewende** (Energy Transition) policy. The country has increased its **renewable energy share from 6% in 2000 to over 45% by 2021**, significantly reducing greenhouse gas emissions (Hake et al., 2021).

A major factor in Germany's success is its **feed-in tariff system**, which guarantees long-term payments for electricity generated from renewables (IEA, 2022). This has encouraged private investment in **solar, wind, and biomass energy**, making clean energy economically viable.

Germany has also **phased out nuclear power**, focusing on a future powered by wind and solar energy. By **2022, all nuclear plants were decommissioned**, further emphasizing the country's commitment to a **sustainable energy future** (Hake et al., 2021).

Public support has been instrumental in this transition. Citizens and cooperatives own nearly **50% of Germany's renewable energy capacity**, ensuring widespread participation in clean energy production (IEA, 2022). This decentralized approach has empowered communities to benefit directly from **solar and wind projects.**

Despite challenges such as **grid stability and reliance on coal during peak demand**, Germany's renewable energy transition offers **valuable lessons for other countries**. By prioritizing **long-term policies, financial incentives, and public engagement**, Germany has successfully shifted toward a cleaner, more sustainable energy system.

Bhutan's Carbon-Negative Status

Bhutan is one of the few countries in the world that is **carbon-negative**, meaning it absorbs more carbon dioxide than it emits. This remarkable achievement results from Bhutan's commitment to **conservation, renewable energy, and sustainable policies** (Wangchuk, 2021).

A major factor contributing to Bhutan's success is its **constitutional mandate** to maintain **at least 60% forest cover** at all times. Currently, forests cover over **70% of Bhutan's land**, serving as a massive carbon sink (United Nations Development Programme [UNDP], 2020). The government actively promotes **reforestation and afforestation programs**, ensuring that deforestation remains low.

Bhutan's energy sector also plays a critical role. Nearly **100% of its electricity comes from hydropower**, reducing reliance on fossil fuels (Wangchuk, 2021). Additionally, Bhutan exports surplus clean energy to neighboring countries, further offsetting emissions.

The government has implemented **strict environmental policies**, including bans on plastic bags, subsidies for electric vehicles, and eco-tourism regulations that limit environmental degradation (UNDP, 2020). Bhutan also measures national progress using **Gross National Happiness (GNH)** rather than GDP, prioritizing **sustainability, well-being, and environmental health**.

Bhutan's success proves that **sustainable development and economic progress can coexist**. By balancing **policy, conservation, and renewable energy**, Bhutan has set an example for countries aiming to achieve carbon neutrality.

Sweden's Waste-to-Energy System

Sweden has revolutionized waste management through its **waste-to-energy (WTE) program**, which converts trash into electricity and heat. Remarkably, Sweden **imports waste from other countries** because it has almost eliminated landfill waste (Sundqvist & Palm, 2021).

A key factor in Sweden's success is its **waste hierarchy approach**, prioritizing **reduction, recycling, and energy recovery**. Only **1% of Sweden's waste ends up in landfills**, while nearly **50% is burned to generate energy** (European Environment Agency [EEA], 2020). The WTE plants produce **electricity for 250,000 homes and heat for 1.5 million people**, reducing dependence on fossil fuels.

Sweden has also implemented **strict recycling policies**. Citizens separate waste into multiple categories, ensuring that materials like **plastics, metals, and paper are efficiently recycled** (Sundqvist & Palm, 2021). The government provides **financial incentives for businesses** that adopt circular economy practices, further reducing waste.

Public education campaigns have played a crucial role in Sweden's waste management success. Schools and community programs emphasize **sustainable consumption and proper recycling habits**, increasing citizen participation (EEA, 2020).

Despite criticism regarding **air pollution from incineration**, Sweden's WTE system remains one of the most effective models for waste reduction. Other nations struggling with landfill overflows can learn from Sweden's **integrated approach to waste management.**

Costa Rica's Renewable Energy Leadership

Costa Rica is a global leader in **renewable energy**, generating over **99% of its electricity** from clean sources such as **hydropower, wind, solar, and geothermal energy** (International Renewable Energy Agency [IRENA], 2021). This achievement highlights how **policy, investment, and natural resources** can support a **sustainable energy transition.**

The country benefits from an **abundant supply of rivers and geothermal activity**, making **hydropower and geothermal plants** highly efficient. In 2015, Costa Rica **operated for 299 consecutive days without using fossil fuels**, a testament to its **clean energy capacity** (Barquero, 2020).

Government policies have played a critical role in this success. Costa Rica has **banned oil exploration and extraction**, prioritizing investments in renewable infrastructure instead (IRENA, 2021). Additionally, the government offers **incentives for solar panel installation** and promotes **electric public transportation** to reduce emissions.

Public support has further driven progress. The country has **invested heavily in environmental education**, and citizens actively support **green initiatives** (Barquero, 2020). As a result, Costa Rica's energy transition has broad national backing.

While challenges remain, such as **energy storage and grid modernization**, Costa Rica's achievements prove that a **nation can successfully transition to renewable energy**. Its model provides valuable insights for countries aiming to **reduce fossil fuel dependence**.

Singapore's Water Sustainability Model

Singapore, despite having **limited natural freshwater resources**, has developed a **world-class water management system**, ensuring a **sustainable water supply** through technology, conservation, and policy (Public Utilities Board [PUB], 2021).

One of Singapore's key innovations is its **NEWater program**, which **recycles wastewater into high-quality drinking water** through **advanced filtration and purification processes**. NEWater meets **40% of Singapore's water demand**, reducing reliance on imported water (Lee, 2020).

Desalination is another crucial component. The country has invested in **energy-efficient desalination plants**, which provide **30% of its water needs** (PUB, 2021). This diversification ensures **water security even during droughts**.

Singapore also promotes **rainwater harvesting** and maintains **extensive reservoirs** that collect and store stormwater. The **Marina Barrage**, a multi-purpose dam, not only prevents flooding but also acts as a major freshwater reservoir (Lee, 2020).

Public engagement is another factor in Singapore's success. The government conducts **education campaigns on water conservation**, encouraging citizens to **reduce water waste** (PUB, 2021). Policies such as **progressive water pricing** further incentivize responsible water use.

By integrating **technology, conservation, and public awareness**, Singapore has **overcome water scarcity** and created a **sustainable urban water model** that other nations facing similar challenges can learn from.

South Korea's Smart Cities Initiative

South Korea is at the forefront of **smart city development**, integrating **technology, sustainability, and urban planning** to create **eco-friendly, efficient, and livable urban environments** (Choi & Park, 2021).

One of the most ambitious projects is **Songdo International Business District**, a **fully planned smart city** that prioritizes **green buildings, energy efficiency, and high-tech infrastructure**. The city uses **automated waste collection systems, solar-powered homes**, and **intelligent transportation networks** to reduce emissions (Korean Smart City Association, 2022).

South Korea has also **incorporated artificial intelligence (AI) and big data** into its urban planning. Cities like **Seoul use AI-driven traffic management** to reduce congestion and **smart grids** to optimize energy consumption (Choi & Park, 2021). These innovations lower **carbon footprints and improve urban resilience**.

Public participation is another key element. South Korea's government encourages **citizens to engage in smart city planning**, ensuring that technological advancements align with community needs (Korean Smart City Association, 2022).

Despite challenges such as **high costs and cybersecurity risks**, South Korea's smart cities serve as a **global model** for sustainable urbanization. Other nations can **learn from its success in integrating technology with environmental sustainability**.

The Netherlands' Bicycle Infrastructure

The Netherlands is renowned for its **bicycle-friendly cities**, which contribute to **reduced carbon emissions, improved public health, and sustainable urban transport** (Harms & Kansen, 2021). With **over 35,000 kilometers of dedicated bike paths**, cycling is a **primary mode of transport** for many Dutch citizens (Pucher & Buehler, 2020).

Government policies have played a crucial role in promoting cycling. Cities like **Amsterdam and Utrecht** have implemented **car-free zones, extensive bike lanes, and bicycle parking facilities**, making cycling **safe and convenient** (Harms & Kansen, 2021). Investments in **bike-friendly infrastructure** ensure that more than **50% of short-distance trips in urban areas** are made by bicycle.

Cycling not only reduces **traffic congestion and air pollution** but also improves **public health** by encouraging physical activity. Research shows that **Dutch cyclists live longer on average**, benefiting from **reduced risks of cardiovascular diseases** (Pucher & Buehler, 2020).

The success of **Dutch cycling policies** demonstrates how sustainable **transportation infrastructure** can enhance urban mobility while minimizing environmental impact. Countries seeking to **reduce reliance on cars** can learn from **the Netherlands' approach to urban planning.**

China's Reforestation Efforts

China has undertaken some of the **largest reforestation projects in the world**, reversing decades of **deforestation and desertification** through **policy-driven afforestation programs** (Zhang et al., 2021).

One of the most significant initiatives is the **Grain for Green Program (GFGP)**, launched in **1999** to restore degraded land. Under this program, **farmers receive subsidies to convert croplands into forests**, improving **soil quality and biodiversity** (Liu & Diamond, 2020).

China has also focused on **urban greening**, with cities like **Beijing and Shanghai increasing green spaces** to combat **air pollution and urban heat effects**. The country has planted over **66 billion trees** in the last two decades, contributing to **global carbon sequestration efforts** (Zhang et al., 2021).

Despite criticisms about **monoculture plantations** and the **long-term sustainability of reforestation efforts**, China's approach has significantly **reduced soil erosion and improved air quality.**

This large-scale project serves as a **valuable case study for nations facing land degradation challenges.**

Sweden's Sustainable Waste Management

Sweden is internationally recognized for its **advanced waste management practices**, focusing on **waste reduction, recycling, and waste-to-energy** solutions. The country boasts one of the **highest recycling rates in the world**, with over **99% of waste being recycled or used for energy recovery** (Swedish Environmental Protection Agency, 2020).

Sweden's success is rooted in its **circular economy model**, which encourages **recycling and repurposing waste materials** rather than sending them to landfills. The country operates a **national waste incineration program**, where **non-recyclable waste** is converted into **energy** for heating homes and generating electricity (Andersson & Bäckström, 2021).

The Swedish government has also invested in **advanced sorting technologies**, making it easier for citizens and industries to separate waste into recyclable categories. The **Producer Responsibility Law** ensures that manufacturers are responsible for **collecting and recycling their products** at the end of their lifecycle, fostering a **circular economy** (Swedish Environmental Protection Agency, 2020).

Additionally, Sweden has achieved remarkable success in **reducing landfill use**. In 2020, less than **1% of Sweden's waste was sent to landfills**, which is a drastic decrease compared to the global average (Andersson & Bäckström, 2021).

Sweden's integrated approach to waste management provides a **comprehensive model** for other nations striving to develop sustainable waste practices. Its combination of **recycling, waste-to-energy technologies, and environmental policies** can help countries reduce their **environmental impact** and contribute to global sustainability efforts.

Bhutan's Gross National Happiness and Environmental Sustainability

Bhutan, a small Himalayan kingdom, has adopted a unique approach to development by prioritizing **Gross National Happiness (GNH)** over **Gross Domestic Product (GDP)**. This philosophy intertwines **economic, social, and environmental sustainability**, positioning Bhutan as a **global leader in sustainable development** (Ura et al., 2020).

Bhutan's **environmental policies** reflect its commitment to sustainability. The country has **legally mandated that 60% of its land must remain forested**, which has helped Bhutan become the world's **only carbon-negative country** (Wangchuk, 2020). Bhutan's forests act as a natural **carbon sink**, absorbing more carbon dioxide than the country emits, contributing to global climate mitigation efforts.

Bhutan also emphasizes **sustainable agriculture**, with the government promoting **organic farming** and reducing the use of harmful chemicals (Ura et al., 2020). This approach not only ensures food security but also supports **biodiversity conservation** and **soil health**.

The country's commitment to **renewable energy** is evident through its reliance on **hydropower** to meet domestic energy needs and export electricity to neighboring countries. Bhutan's **hydropower plants** generate **clean energy**, reducing the carbon footprint of its energy sector (Wangchuk, 2020).

Bhutan's GNH framework provides an alternative model to traditional development, emphasizing the **importance of well-being, environmental stewardship**, and **sustainable growth**. Other countries can learn from Bhutan's holistic approach to integrating environmental conservation with **social and economic policies**.

Norway's Electric Vehicle Transition

Norway is leading the **global shift towards electric vehicles (EVs)**, with a remarkable **over 50% of all new cars sold** in the country being **fully electric** (Norwegian Road Federation, 2020). This transformation is part of a larger commitment to **reduce carbon emissions** and **promote green transportation solutions.**

One of the key drivers of this success is the government's **generous incentives for EV buyers**, including **exemptions from VAT and road tolls, free parking**, and **access to bus lanes** (Andersen, 2021). These policies have made electric vehicles financially attractive to Norwegian citizens, increasing the adoption rate of EVs faster than in any other country.

Norway's government has set an ambitious goal of **phasing out the sale of petrol and diesel cars by 2025**, further incentivizing the transition to **clean vehicles** (Norwegian Road Federation, 2020). The country has also invested heavily in the **EV charging infrastructure**, ensuring that **charging stations** are accessible across urban and rural areas.

The success of Norway's **electric vehicle transition** offers valuable lessons for other nations seeking to reduce **transportation emissions.** By providing strong financial incentives, improving infrastructure, and setting clear policy goals, Norway has created a **comprehensive strategy** for **electrifying the transport sector**.

Kenya's Solar Energy Expansion

Kenya is making significant strides in the **adoption of solar energy**, especially in **rural areas**, where **access to electricity** has traditionally been limited. Solar power is playing an essential role in providing **clean and affordable energy** to communities without reliable access to the national grid (Kenya Renewable Energy Association [KREA], 2021).

One of Kenya's most notable initiatives is the **M-KOPA Solar program**, which enables families to purchase **solar home systems** on a **pay-as-you-go basis** through mobile money. This system has empowered over **1 million households** to access **solar power**, significantly reducing reliance on **kerosene lamps** (Ochieng, 2020).

Kenya's government has also invested in large-scale solar power projects, including **the Garissa Solar Plant**, which provides **clean energy** to the national grid (KREA, 2021). Additionally, Kenya is expanding its **solar energy infrastructure** to support the increasing demand for electricity as the population grows.

Solar energy adoption has reduced **carbon emissions** and improved **quality of life** by providing a more **reliable and sustainable energy source**. Kenya's experience underscores the potential for **solar power** to improve energy access and **combat climate change** in developing nations.

Costa Rica's Renewable Energy Leadership

Costa Rica is recognized globally for its commitment to **renewable energy**, with the country meeting nearly **100% of its electricity needs from renewable sources** (Vargas & Sánchez, 2021). The country's energy portfolio includes **hydropower, wind, solar, and geothermal energy**, with a heavy reliance on **hydropower**, which accounts for more than **70%** of the nation's electricity generation (Solis et al., 2020).

One of the key factors behind Costa Rica's success is its **investment in clean energy infrastructure** and the development of **geothermal energy plants** in volcanic regions. The government has created a favorable policy environment by incentivizing **private sector involvement** in renewable energy projects and promoting **energy efficiency** (Solis et al., 2020).

Costa Rica has also set ambitious targets for achieving **carbon neutrality** by 2050. As part of its strategy, the country has implemented **reforestation programs**, invested in **electric public transportation**, and promoted **sustainable agriculture practices** (Vargas & Sánchez, 2021). These actions contribute to **reducing**

greenhouse gas emissions and enhancing the country's environmental sustainability.

The country's **proactive approach** to renewable energy has not only contributed to **reducing environmental impact** but has also created **economic opportunities** in green technology and job creation in the renewable energy sector. Costa Rica's model provides valuable insights into how countries can shift to **sustainable energy systems** and achieve **long-term environmental and economic goals**.

The Role of Community Gardens in Urban Sustainability

Community gardens have become a significant part of **urban sustainability efforts**, particularly in cities seeking to enhance **food security**, **reduce food miles**, and **promote environmental education**. These gardens allow urban residents to grow their own food while contributing to **biodiversity** and reducing **carbon emissions** associated with food transportation (Kaufman & Firth, 2021).

In cities like **New York**, **Detroit**, and **London**, community gardens have become a tool for **building social cohesion** and **empowering marginalized communities**. These gardens often occupy vacant lots and transform underused spaces into productive, green areas that provide **fresh produce** and **green spaces** for local residents (Gibson & Matthews, 2020).

Community gardens also serve as **educational hubs**, teaching participants about **sustainable farming practices**, **composting**, and **resource conservation**. Many urban gardens incorporate **rainwater harvesting systems** and **composting programs**, which help reduce water waste and promote sustainable waste management (Kaufman & Firth, 2021).

Furthermore, community gardens help mitigate the effects of **urban heat islands** by providing green cover that **absorbs heat** and

improves air quality. They also contribute to **mental well-being** by offering a space for relaxation and community engagement.

The rise of **urban agriculture**, particularly through community gardens, illustrates the potential for cities to improve **food systems, environmental sustainability**, and **community resilience**. By embracing this model, cities can enhance their **green credentials** and foster a **sustainable urban environment**.

The Solar Energy Revolution in Germany

Germany has emerged as a leader in the global **solar energy revolution**, with one of the **largest solar power installations** in the world. Through the **Energiewende** (energy transition) initiative, the country has committed to reducing carbon emissions while promoting **renewable energy sources**, including **solar** (Frondel et al., 2021).

The success of Germany's **solar power expansion** lies in its **innovative policies**, such as the **Feed-in Tariff** (FiT), which guarantees long-term contracts for renewable energy producers at favorable rates. These incentives have encouraged both private citizens and corporations to invest in **solar energy systems**, propelling Germany to install over **50 GW of solar capacity** (Renn, 2020).

Germany's **solar energy expansion** has had numerous positive outcomes, including the creation of **thousands of jobs** in the green energy sector and a significant reduction in **carbon emissions**. By 2020, solar power contributed to over **10% of Germany's electricity needs**, making it a pivotal part of the country's energy mix (Frondel et al., 2021).

Moreover, Germany's commitment to **sustainability** has set a precedent for other nations looking to scale up their renewable energy sectors. The combination of **strong government policies, public-private partnerships**, and **technological innovation** has made Germany a **global role model** for solar energy adoption.

The Success of the Green Building Movement in Australia

Australia has made significant strides in **green building** initiatives, largely driven by the **Green Building Council of Australia (GBCA)** and its **Green Star** rating system. This movement focuses on creating **energy-efficient, sustainable buildings** that reduce environmental impact while improving the quality of life for their occupants (Royston et al., 2020).

A prime example of success is the **One Central Park** project in Sydney, which features **solar power, water recycling, and green spaces**. This building achieved a **6-Star Green Star rating**, demonstrating the high potential of **integrating sustainable practices** into commercial real estate developments. The **solar energy systems** installed on the building's rooftops provide significant energy savings and reduce the **carbon footprint** (GBCA, 2019).

The **Green Star** certification has become a standard in Australian construction, incentivizing developers to adopt **sustainable materials**, improve **energy efficiency**, and enhance **water conservation**. As of 2021, over **600 projects** across the country have received **Green Star** certifications, reflecting the growing demand for **sustainable buildings** (Royston et al., 2020).

This movement is also beneficial for businesses, as **green buildings** tend to attract **higher rents**, **lower operational costs**, and greater **market appeal**. By adopting environmentally friendly practices, Australia has set a benchmark for **sustainable urban development**.

Japan's Environmental Education Programs

Japan's approach to **environmental education** has been instrumental in raising public awareness about sustainability. With a long history of environmental activism, the country has developed a variety of **education programs** that engage both young and adult populations in environmental conservation (Sato, 2021).

One notable example is the **Environmental Education Promotion Act** passed in 2003, which mandated the incorporation of environmental topics into school curricula. Through **hands-on activities** and **experiential learning**, students are encouraged to understand the relationship between **human activities** and the **natural environment** (Sato, 2021).

Japan has also implemented **community-based environmental education programs**, where local residents are taught sustainable practices like **recycling**, **composting**, and **energy conservation**. For instance, the city of **Kobe** has seen a dramatic reduction in waste sent to landfills due to educational initiatives that promote **zero waste** living (Yamamoto et al., 2020).

Furthermore, Japan's **environmental education programs** have extended to corporate training, with many companies offering **sustainability workshops** for employees. These programs foster a culture of **environmental responsibility** within both the **workplace** and **wider community**.

By emphasizing education at all levels of society, Japan has successfully created a culture of **sustainability** that informs policy, personal behavior, and community development.

The Circular Economy in the Netherlands

The Netherlands has positioned itself as a leader in the adoption of the **circular economy**, where the emphasis is on reducing waste and maximizing the use of **recycled materials**. This shift from a **linear economy** to a **circular economy** is part of the Netherlands' broader strategy to **combat climate change** and reduce environmental harm (Circular Economy Netherlands, 2020).

One of the most successful initiatives is the **Dutch Waste to Energy Program**, which processes waste materials and turns them into **useful resources**, including **energy**, **compost**, and **recycled**

products. This initiative reduces **landfill use** and ensures that valuable materials like **plastics, metals, and paper** are recycled back into the economy (Circular Economy Netherlands, 2020).

The government's **circular economy policies** have made recycling and reusing materials part of daily life. One example is the widespread use of **deposit return schemes** for plastic bottles and aluminum cans, which encourage people to **return used containers** for recycling (de Lange, 2021).

By promoting the idea of a **closed-loop system**, the Netherlands has been able to significantly reduce its **waste production** while creating economic value. The country's success in the **circular economy** is setting an example for other nations that seek to address the environmental challenges posed by **waste management** and **resource depletion.**

The Rise of Electric Vehicles in Norway

Norway has become a **global leader** in the adoption of **electric vehicles (EVs)**, with nearly **50%** of all new car sales in the country being **electric** (Horne & Kåre, 2021). This success can be attributed to a combination of **government incentives**, **environmental awareness**, and the development of an extensive **charging infrastructure** that has made electric vehicles a **mainstream option** for Norwegian consumers.

The Norwegian government offers a range of benefits to those purchasing electric vehicles, including **tax exemptions**, **free tolls**, and **access to bus lanes**. These incentives have made EVs more affordable and attractive compared to traditional gasoline-powered cars. In addition, the government has set ambitious targets to reduce **carbon emissions** and to transition toward a **carbon-neutral society** by 2050 (Horne & Kåre, 2021).

Furthermore, Norway's efforts to expand its **EV charging infrastructure** have played a key role in supporting the widespread adoption of electric cars. There are now thousands of **public**

charging stations across the country, making it convenient for drivers to charge their EVs while on the go (Johansen et al., 2020).

Norway's success in promoting **electric vehicles** offers a valuable case study for other countries looking to reduce their **carbon footprint** and transition to **clean energy solutions**. The country's efforts have been instrumental in encouraging the global adoption of **electric transportation**, paving the way for a more sustainable future.

The Role of Reforestation in China's Environmental Recovery

China has implemented one of the **largest reforestation programs** in the world, aimed at combating **deforestation** and improving **environmental conditions** across the country. Over the past several decades, China has planted **billions of trees** to restore ecosystems, improve air quality, and mitigate the impacts of **climate change** (Li et al., 2020).

One of the most notable projects is the **Green Great Wall** initiative, which aims to combat desertification by planting trees along the northern borders of China. This massive project covers over **100 million hectares** of land and has significantly improved soil quality, water retention, and **biodiversity** in the affected areas (Li et al., 2020).

Reforestation efforts have not only contributed to environmental recovery but also provided significant **economic benefits**. The planting of trees creates jobs in rural areas, boosts the timber industry, and attracts ecotourism. These programs have become a source of **economic growth** while simultaneously addressing **environmental degradation** (Xu et al., 2021).

China's reforestation success highlights the potential of **large-scale ecological restoration projects** in combating environmental issues and improving long-term sustainability. The country's efforts show that **government policy, national**

commitment, and **local action** can collectively drive significant positive change in environmental recovery.

The Success of Ecotourism in Costa Rica

Costa Rica has become a **global leader** in **ecotourism**, demonstrating how **sustainable tourism practices** can contribute to both **economic development** and **environmental conservation**. The country's rich biodiversity, diverse ecosystems, and commitment to **preserving natural habitats** have made it a top destination for eco-conscious travelers (Bennett & Gough, 2020).

The government of Costa Rica has implemented a series of policies that encourage **sustainable tourism**, including the **Payment for Environmental Services (PES)** program, which compensates landowners for **conserving** and **protecting forests**. This program has led to the preservation of over **25% of the country's land area** as protected national parks and reserves (Bennett & Gough, 2020).

In addition to the environmental benefits, **ecotourism** has become a significant economic driver for Costa Rica, providing **jobs**, supporting **local businesses**, and promoting a **green economy**. The country's focus on ecotourism has also educated both visitors and locals about the importance of **biodiversity conservation** and **sustainable resource management** (Klein, 2021).

Costa Rica's success in developing an ecotourism model underscores the potential for other countries to use **sustainable tourism** as a tool for both **economic growth** and **environmental protection**. By creating a harmonious balance between conservation and economic development, Costa Rica provides a blueprint for **sustainable development** in the tourism industry.

Conclusion: A vision for a sustainable future

Introduction to Sustainability

Sustainability has become one of the most critical global issues of the 21st century. As the world faces growing challenges such as **climate change**, **biodiversity loss**, and **resource depletion**, it is crucial that societies adopt sustainable practices to ensure a viable future for generations to come. The concept of sustainability encompasses three interconnected pillars: **environmental**, **social**, and **economic** (Daly & Farley, 2018).

To achieve sustainability, it is essential that all sectors of society—governments, businesses, and individuals—play an active role in driving change. Governments must create policies that promote **renewable energy**, **conservation efforts**, and **green technologies** while ensuring that economic growth does not come at the cost of the environment (Elkington, 1997). At the same time, businesses must integrate sustainability into their operations by adopting **eco-friendly practices** and **corporate social responsibility** (CSR) initiatives (Porter & Kramer, 2011).

Individual actions, such as reducing waste, conserving water, and supporting sustainable products, also contribute significantly to **sustainable living**. Public awareness and education are paramount in fostering a sense of responsibility toward the planet and its resources. The journey toward sustainability requires a collective commitment to create lasting change, and every effort, no matter how small, adds up toward building a better future.

Ultimately, the vision for a **sustainable future** is one in which society thrives in harmony with the planet, fostering ecological integrity, economic equity, and social well-being for all (Meadows et al., 2004).

The Importance of Renewable Energy

A critical aspect of achieving a **sustainable future** is the transition to **renewable energy** sources such as **solar**, **wind**, and **hydropower**. Fossil fuels, which have powered industrial growth for centuries, are a major driver of **climate change** due to their contribution to **greenhouse gas emissions**. The global shift toward renewable energy aims to reduce reliance on these polluting energy sources and transition toward cleaner, more sustainable alternatives (IEA, 2020).

Solar energy, for example, has seen significant growth in both **developed** and **developing** countries. Technological advancements in solar panels, as well as falling costs, have made solar power more accessible and efficient. Likewise, **wind energy** has emerged as a **powerful tool** in reducing carbon emissions, with offshore wind farms proving particularly effective in high-wind areas (Lazard, 2020).

Governments, businesses, and communities must collaborate to **increase investment** in **renewable energy infrastructure** while promoting policies that encourage **energy efficiency**. With the right incentives, renewable energy has the potential to power entire nations sustainably. The future of energy is renewable, and embracing these technologies is essential for achieving a more sustainable, resilient world.

Additionally, investing in **clean energy technologies** will create jobs and stimulate economic growth while ensuring the preservation of natural ecosystems. The vision of a sustainable future is one in which **renewable energy** is the **dominant source of power**, paving the way for **cleaner air**, **reduced greenhouse gas emissions**, and a **more equitable energy distribution** across the globe (World Bank, 2020).

Conservation and Biodiversity Protection

A vision for a sustainable future cannot be realized without a robust commitment to **conservation** and **biodiversity protection**.

Healthy ecosystems provide essential services such as clean air, water, and soil, all of which are crucial for human survival. Yet, the ongoing **loss of biodiversity** due to **deforestation**, **pollution**, and **climate change** poses significant threats to both nature and humanity (Rockströḿ et al., 2009).

To preserve biodiversity, it is essential to implement effective **conservation strategies** that protect both land and marine ecosystems. For instance, establishing **protected areas**, such as **national parks** and **marine reserves**, has been a key strategy for maintaining biodiversity and allowing ecosystems to thrive. In addition to conserving natural areas, there is a growing need for **restoration efforts** to regenerate ecosystems that have been damaged by human activity (Sala et al., 2000).

Sustainable practices in agriculture, forestry, and fisheries are also crucial for maintaining biodiversity. **Agroecology** and **sustainable farming** techniques that prioritize the health of the land and soil can ensure that food production does not come at the expense of the environment. Similarly, promoting **sustainable fisheries** and reducing overfishing will help protect marine biodiversity and ensure that ocean ecosystems remain balanced.

In a sustainable future, protecting biodiversity is a **shared responsibility** that requires coordinated action at local, national, and global levels. By working together to **conserve natural habitats** and restore ecosystems, humanity can create a world where both people and nature can thrive (TEEB, 2010).

Sustainable Agriculture and Food Systems

One of the most significant challenges in building a sustainable future is ensuring food security while minimizing the environmental impact of agriculture. **Sustainable agriculture** focuses on producing sufficient food to meet the needs of a growing global population while promoting environmental conservation and social equity (Gliessman, 2016).

Agroecology is a promising approach to sustainable agriculture, emphasizing the integration of ecological principles into farming systems. This method encourages farmers to adopt practices such as **crop diversification**, **organic farming**, and **soil health management**, which improve productivity while minimizing the use of harmful chemicals and reducing environmental degradation.

Regenerative agriculture also plays a crucial role in reversing environmental damage caused by conventional farming practices. By focusing on soil health, **carbon sequestration**, and the restoration of natural ecosystems, regenerative agriculture has the potential to reduce **carbon emissions**, improve **water retention**, and enhance **biodiversity** (Anderson, 2019).

In a sustainable future, food systems must prioritize **local production**, **fair trade**, and **food sovereignty**, ensuring that communities have access to healthy, affordable food while reducing dependence on global supply chains that often contribute to **food waste** and **environmental harm**. Transitioning to sustainable agriculture is essential to feeding the world in a way that is both **environmentally responsible** and **socially just** (FAO, 2020).

Green Technologies and Innovation

The role of **green technologies** is pivotal in shaping a sustainable future. These technologies are designed to mitigate environmental harm while promoting the efficient use of resources. Innovations such as **renewable energy systems, energy-efficient buildings**, and **clean transportation** are all contributing to the shift toward a low-carbon, sustainable economy. Green technologies are not only essential for reducing **greenhouse gas emissions** but also for fostering **economic growth** through the creation of new industries and jobs (Sachs, 2015).

For instance, **solar energy** technology has witnessed tremendous advancements, enabling more efficient **solar panels** with lower costs. Likewise, **wind turbines** and **geothermal energy** are providing clean alternatives to fossil fuel-based energy. These renewable energy technologies are crucial in the global transition to a carbon-neutral future (Jacobson et al., 2015).

Moreover, the development of **electric vehicles** (EVs) and **battery storage systems** holds great potential for reducing reliance on fossil fuels in the transportation sector. EVs have become increasingly popular due to advancements in **battery technology**, which have improved the range and affordability of electric cars. Coupled with a growing network of **charging infrastructure**, EVs are becoming a viable alternative to gasoline-powered vehicles (Breetz et al., 2018).

Green technologies not only benefit the environment but also stimulate innovation and create a more resilient, sustainable economy. The future of these technologies lies in their ability to integrate with existing infrastructures while promoting a shift toward more sustainable practices across industries and society (Schwab, 2016).

Circular Economy and Resource Efficiency

The concept of a **circular economy** is increasingly recognized as a key component of sustainable development. Unlike the traditional **linear economy**, where resources are extracted, used, and discarded, the circular economy focuses on **resource efficiency**, **waste reduction**, and **recycling**. This model encourages the design of products that can be reused, repaired, and recycled, reducing the need for raw material extraction and minimizing waste production (Geissdoerfer et al., 2017).

One of the core principles of the circular economy is **design for longevity**, where products are designed with durability and **end-of-life recyclability** in mind. By designing for longevity, companies can reduce the environmental impact of products and decrease the frequency with which new products need to be manufactured. This reduces both **energy consumption** and **waste**

generation, contributing to a more sustainable economy (EMF, 2015).

Furthermore, **industrial symbiosis** plays a crucial role in this model by encouraging collaboration between different industries to maximize the use of by-products and waste. This leads to **closed-loop systems**, where the waste from one company becomes the input for another, reducing the need for virgin materials. Many businesses and cities around the world are adopting circular practices, thereby creating new economic opportunities while improving environmental outcomes (Lewandowski, 2016).

The transition to a circular economy requires a fundamental shift in both business practices and consumer behavior. Governments can support this transition through policies that incentivize **sustainable design**, **waste reduction**, and **recycling programs**. By embracing the principles of a circular economy, societies can move toward a future where resources are used efficiently and waste is minimized.

Education and Awareness for Sustainability

Achieving a sustainable future is not solely reliant on technological advancements and policy changes; it also requires a fundamental shift in societal values and attitudes. **Education and awareness** are essential tools in empowering individuals, communities, and organizations to make more informed, sustainable choices. In many cases, people are unaware of the environmental consequences of their actions, which hinders efforts toward sustainability (UNESCO, 2017).

By integrating **environmental education** into school curricula and **community outreach programs**, individuals can develop a deeper understanding of the pressing environmental challenges we face and the solutions available. Such education fosters a sense of **responsibility** and encourages people to adopt sustainable behaviors, such as reducing energy consumption, supporting eco-friendly products, and reducing waste. The impact of educational programs can be far-reaching, creating a ripple effect as individuals share their knowledge with others, thus promoting widespread change.

Further, **public awareness campaigns** can raise consciousness about the importance of sustainability. Governments, nonprofits, and businesses have increasingly used social media and other platforms to engage the public in discussions about environmental issues. These platforms allow for a broader dissemination of information and can encourage people to participate in environmental initiatives such as **tree planting**, **recycling drives**, and **clean-up events** (Liu et al., 2020).

Investing in education and raising awareness is integral to achieving a sustainable future, as it empowers individuals to make choices that prioritize the well-being of both people and the planet. Through informed actions, communities can collectively contribute to a more sustainable world.

The Role of Governments in Sustainability

Governments play a critical role in achieving sustainability by establishing policies that promote **environmental protection**, **economic equity**, and **social justice**. The implementation of effective environmental laws and regulations is essential to the transition to a sustainable future. This includes **carbon pricing**, regulations on **pollution control**, and incentives for renewable energy adoption (Stern, 2007).

International cooperation is also vital, as many environmental challenges, such as **climate change**, **biodiversity loss**, and **ocean pollution**, are global in nature. Governments must collaborate across borders to address these issues and share knowledge and resources for sustainable development (UNFCCC, 2015). Policies such as the **Paris Agreement**, which commits countries to limit global temperature rise, are key examples of multilateral efforts to combat climate change.

National governments must also incentivize **green innovation** through financial support, research grants, and public-private partnerships. By investing in clean technologies and sustainable practices, governments can foster the creation of **green jobs** and help stimulate the economy while mitigating environmental damage. Policies that promote **sustainable agriculture**, **conservation**, and **biodiversity protection** are also essential for preserving natural resources for future generations (World Bank, 2020).

The actions of governments can set the stage for broader societal shifts toward sustainability. By enacting forward-thinking policies, governments can create the regulatory framework necessary to ensure that businesses and individuals align their actions with sustainability goals.

Technological Innovations and Their Impact on Sustainability

Technological innovation has always played a crucial role in the advancement of society, and it will continue to be essential in achieving a sustainable future. Innovations in **clean energy**, **water purification**, and **waste management** technologies are particularly important in addressing the environmental challenges we face. The development of **smart grids**, **solar panels**, and **battery storage** is transforming the way energy is produced, consumed, and stored, providing cleaner alternatives to fossil fuels (Feldman et al., 2019).

For instance, advancements in **solar photovoltaic (PV) technology** have drastically reduced the cost of solar power, making it a viable energy source for more people worldwide. According to a study by the International Renewable Energy Agency (IRENA), the cost of solar energy has fallen by more than 80% in the last decade, leading to increased adoption of solar systems across the globe (IRENA, 2020).

Likewise, innovations in **electric vehicles** and **battery technologies** are expected to play a major role in decarbonizing the transportation sector. As EVs become more affordable and accessible, the world will see a reduction in dependence on **gasoline-powered cars**, reducing **carbon emissions** and **air pollution** (Breetz et al., 2018).

These technological advancements, combined with **sustainability-focused policies**, can drive the transition to a low-carbon economy. By continuing to innovate and invest in green technologies, we can achieve a future where **renewable energy** powers our cities, **electric vehicles** replace gas-powered cars, and **smart technologies** help us better manage our resources.

Sustainable Consumption and Production

The concept of **sustainable consumption and production** (SCP) refers to the need to promote resource efficiency and minimize environmental harm through better production and consumption patterns. This involves reducing waste, improving the efficiency of resource use, and ensuring that production methods are socially responsible and environmentally friendly (UNEP, 2018). SCP is closely aligned with the **circular economy**, which seeks to ensure that resources are used in a way that can be **recycled** or **reused** to minimize waste and pollution.

A shift toward sustainable consumption requires a change in consumer behavior. This could involve choosing **eco-friendly products**, **supporting local and sustainable businesses**, and adopting more sustainable practices in daily life. It also entails greater emphasis on the **life cycle of products**—from production to disposal—ensuring that products are designed with their environmental impact in mind (Bocken et al., 2016).

Similarly, sustainable production focuses on encouraging businesses to implement **greener** and **more efficient manufacturing processes**. Companies are increasingly adopting **clean technologies**, **energy-efficient equipment**, and **low-carbon materials** in their production lines to minimize environmental harm. These practices not only help reduce emissions but also contribute to the long-term sustainability of the supply chain.

The promotion of SCP involves collaboration among governments, industries, and consumers. Policies that encourage resource efficiency, eco-design, and waste reduction can incentivize businesses to adopt greener practices, while public awareness campaigns can drive consumer demand for sustainable products.

International Cooperation and Global Initiatives

Addressing global environmental challenges requires a unified approach from governments, organizations, and individuals worldwide. Many environmental issues, such as **climate change**, **biodiversity loss**, and **pollution**, are transboundary in nature, meaning they cannot be effectively tackled by one country or entity alone. **International cooperation** is therefore essential to create comprehensive solutions and share resources, knowledge, and best practices.

International agreements like the **Paris Agreement** on climate change, the **Convention on Biological Diversity**, and the **United Nations 2030 Agenda for Sustainable Development** are key milestones in the global effort to achieve sustainability. These agreements bring countries together to set common goals and targets, with the aim of reducing global carbon emissions, protecting ecosystems, and promoting sustainable development (UNFCCC, 2015; UNDP, 2020).

Through multilateral cooperation, nations can share **green technologies**, fund **sustainable development projects** in lower-income countries, and ensure that all countries have access to the tools and knowledge needed to address environmental challenges. Global financial mechanisms, such as the **Green Climate Fund** and the **Global Environment Facility**, play a key role in supporting these efforts by providing funding for sustainable development initiatives (World Bank, 2020).

Furthermore, international collaboration can foster the creation of **global supply chains** that prioritize sustainability and ensure that all stakeholders work together to meet the **Sustainable Development Goals** (SDGs). As the world becomes more interconnected, it is critical for countries to support each other in their efforts toward sustainability.

The Role of Businesses in Sustainability

In recent years, businesses have increasingly recognized their role in promoting sustainability. Rather than focusing solely on profits, many companies are now adopting **sustainable business practices** that prioritize environmental protection, resource conservation, and social responsibility. These efforts are driven by consumer demand for **eco-friendly products**, **investor expectations**, and an increasing awareness of the environmental impacts of business activities.

Corporate social responsibility (CSR) initiatives and the **integration of sustainability into business strategies** have become central to many organizations. Businesses are focusing on reducing their **carbon footprint**, improving **energy efficiency**, and **reducing waste** through **recycling programs** and **product redesigns**. Companies in various sectors, from **fashion** to **tech**, are incorporating **sustainable sourcing** into their supply chains and aiming to reduce the impact of their products and services on the environment (Porter & Kramer, 2011).

Moreover, many companies are adopting **green certifications** and **sustainability reporting** to demonstrate their commitment to environmental and social causes. These efforts not only benefit the environment but also enhance the company's reputation, which can lead to increased customer loyalty, higher sales, and access to sustainable investment opportunities.

By adopting sustainable practices, businesses play an instrumental role in driving the **green economy** and fostering the transition toward a more sustainable global economy. Their actions, when coupled with government policies and public awareness, will help create a sustainable future for all.

The Role of Education in Sustainability

Education plays a critical role in shaping the attitudes, knowledge, and behaviors of individuals and communities in relation to sustainability. By fostering a deep understanding of environmental issues and the importance of sustainable practices, educational initiatives can equip future generations with the tools needed to create lasting change. Schools, universities, and community organizations all have the potential to cultivate a sense of responsibility for the environment through curriculum development, outreach programs, and awareness campaigns (Tilbury, 2017).

Environmental education emphasizes the interconnectedness of environmental, social, and economic issues. By promoting concepts such as **sustainable development**, **resource conservation**, and **climate change awareness**, it empowers individuals to make informed decisions about their actions and consumption. In schools, students learn about topics ranging from **biodiversity** to **green technologies**, providing them with the knowledge to pursue careers in fields like renewable energy, environmental science, and sustainable agriculture (UNESCO, 2014).

Higher education institutions also play a pivotal role in advancing sustainability research and innovation. Many universities have established **sustainability research centers** that focus on developing new technologies and solutions to environmental challenges. Additionally, **professional development programs** that offer training in sustainability practices for working professionals can further promote environmentally conscious behaviors across industries (Barton, 2017).

Moreover, **public awareness campaigns** supported by educational institutions and non-governmental organizations

(NGOs) help to spread crucial information about sustainable lifestyles. By making sustainability a part of everyday learning, education can cultivate an informed society ready to tackle environmental challenges head-on.

The Importance of Conservation in Sustainability

Conservation efforts are at the core of any sustainability agenda. Whether it is conserving **natural resources**, **biodiversity**, or **ecosystems**, these actions help preserve the planet's health and ensure that future generations can enjoy a livable environment. Conservation involves protecting and maintaining the **natural world** in a way that allows humans to thrive while reducing **ecological damage** and avoiding depletion of finite resources.

One of the most significant areas of conservation is **biodiversity conservation**. The protection of various species, ecosystems, and genetic resources is essential for the stability of natural processes such as **pollination**, **water purification**, and **soil regeneration**. Without a healthy and diverse range of species, ecosystems would collapse, leading to reduced agricultural yields, water shortages, and the loss of essential ecological services (Cardinale et al., 2012).

Similarly, **water conservation** plays a critical role in ensuring sustainable development. Freshwater resources are limited, and with growing populations and climate change, many regions face water scarcity. Implementing water-saving techniques in agriculture, industry, and daily life can significantly reduce water consumption and mitigate the effects of droughts and flooding (Postel, 2000).

In addition to these efforts, **forest conservation** is vital. Forests play an essential role in sequestering carbon, regulating climate, and providing habitat for wildlife. Protecting forests from **deforestation** is an essential step in mitigating **climate change** and preserving biodiversity.

Effective conservation requires both **policy action** and **community involvement**. Governments must implement and enforce protective laws and regulations, while individuals and local communities must adopt sustainable practices to reduce their environmental footprint.

Renewable Energy and Sustainable Practices

The transition to **renewable energy** is a crucial step toward achieving sustainability and addressing climate change. Fossil fuels such as coal, oil, and natural gas are major contributors to **greenhouse gas emissions**, which are responsible for global warming and environmental degradation. By shifting to renewable energy sources such as **solar**, **wind**, **hydropower**, and **geothermal** energy, we can reduce our reliance on polluting fossil fuels and decrease our environmental impact.

Renewable energy offers multiple environmental and economic benefits. For instance, **solar energy** harnesses the power of the sun to generate electricity, and it produces no direct emissions. The price of solar panels has significantly dropped in recent years, making solar energy more accessible and affordable for businesses and households (IRENA, 2020). Similarly, **wind energy** is abundant and has minimal environmental impact. Offshore wind farms, in particular, offer vast potential for generating electricity in areas where land-based wind resources are limited (Lund, 2019).

Wind and solar power are becoming increasingly integrated into national and international energy grids, reducing the need for coal and gas-powered plants. This transition has the potential to dramatically reduce global **carbon emissions**, contributing to a cleaner, more sustainable energy future (Jacobson et al., 2017).

Furthermore, the adoption of renewable energy is essential for addressing **energy access** in developing regions. Off-grid solar power, for instance, can provide electricity to rural communities that have historically lacked access to centralized power grids, improving quality of life and enabling economic development (UNDP, 2018).

The Need for Policy Action in Sustainability

Policies play a central role in shaping the trajectory of sustainability efforts. Governments worldwide are beginning to recognize the importance of policies that promote **green growth**, **environmental protection**, and **resource conservation**. However, achieving sustainability at the global scale requires a comprehensive policy approach that involves collaboration at the local, national, and international levels.

Environmental policies must be designed to address the root causes of environmental degradation. These policies should include incentives for businesses to adopt **sustainable practices**, regulations to curb **carbon emissions**, and support for **clean technologies** and **renewable energy** (Stern, 2007). For example, governments can introduce **carbon pricing mechanisms** or **tax incentives** to encourage the use of clean energy and energy-efficient technologies. Similarly, policies that promote **sustainable land use** can help prevent **deforestation** and encourage the restoration of damaged ecosystems (FAO, 2019).

At the international level, policies that promote **global environmental cooperation** are essential for tackling issues such as climate change, biodiversity loss, and resource depletion. Agreements such as the **Paris Agreement** have set global targets for emissions reductions, and many countries have committed to achieving **net-zero emissions** by mid-century (UNFCCC, 2015).

Governments also play a key role in promoting **public awareness** about environmental issues. Educational campaigns that inform citizens about the importance of sustainability can encourage individuals to make eco-friendly choices, from reducing energy consumption to adopting plant-based diets.

Ultimately, **effective policies** are those that **empower communities**, **support innovation**, and incentivize individuals and businesses to act in ways that benefit both people and the planet.

Economic Incentives for Sustainable Development

Economic incentives are powerful tools in promoting sustainability across sectors. Governments and organizations can leverage various mechanisms to encourage businesses and individuals to adopt sustainable practices. These incentives can include **tax credits**, **subsidies**, **grants**, and **reduced tariffs** for products or activities that have a lower environmental impact. For example, **solar panel installations** are often subsidized to make renewable energy more affordable and accessible to consumers, promoting a cleaner energy future (Ren21, 2020).

Additionally, **carbon pricing** mechanisms, such as carbon taxes and **cap-and-trade systems**, aim to put a price on carbon emissions, encouraging companies to reduce their emissions and invest in cleaner technologies. This incentivizes businesses to internalize the environmental costs of their activities, rather than passing those costs onto society in the form of pollution and climate change impacts (Harris, 2016).

Another incentive that has gained attention is **green bonds**, which are financial instruments used to fund environmentally sustainable projects, such as renewable energy development or energy-efficient infrastructure. Green bonds enable businesses and governments to raise capital for sustainability initiatives while allowing investors to support eco-friendly projects (Flammer, 2021).

Furthermore, **corporate social responsibility (CSR)** programs also provide an incentive for businesses to act sustainably. By aligning business operations with environmental and social goals, companies can enhance their brand image, attract environmentally conscious consumers, and improve employee satisfaction, which can ultimately result in greater profitability in the long run (Morgensen, 2018).

Ultimately, the careful design and implementation of economic incentives can create a powerful system of rewards for those who prioritize sustainable practices, driving large-scale environmental change.

Technological Innovations for Sustainability

The role of **technology** in driving sustainability cannot be overstated. Innovations in fields like **renewable energy**, **resource management**, and **sustainable agriculture** are transforming the way we produce, consume, and dispose of resources. **Green technologies** are helping to minimize waste, reduce carbon footprints, and increase the efficiency of energy use across industries.

Energy storage technologies have made significant progress in recent years, addressing one of the biggest challenges in renewable energy: intermittency. Advances in battery storage, such as lithium-ion and solid-state batteries, allow energy generated from **solar** and **wind** sources to be stored and used when demand is high, improving the reliability of renewable energy systems (Dunn et al., 2019). Additionally, **smart grids** allow for real-time energy distribution, optimizing the use of renewable resources and improving the efficiency of the power grid.

In **agriculture**, **precision farming technologies** are helping to reduce waste and increase crop yields through data-driven decision-making. Techniques like **drone-assisted monitoring**, **GPS-guided planting**, and **automated irrigation systems** allow farmers to optimize the use of water, fertilizers, and pesticides, leading to more sustainable practices and reduced environmental impact (Zhang et al., 2020).

Moreover, the rise of **circular economy models** is contributing to sustainability by encouraging the reuse and recycling of materials, rather than their disposal. Technologies that facilitate **waste-to-energy processes**, such as anaerobic digestion and **advanced recycling technologies**, help reduce landfill waste while generating usable energy (Ellen MacArthur Foundation, 2019).

As we continue to innovate, technology remains a key enabler in achieving a more sustainable world.

Collaborative Efforts and Global Partnerships

No single nation, organization, or individual can solve the world's environmental challenges alone. Global issues such as **climate change, biodiversity loss**, and **resource depletion** require **collaborative action** at every level—from local communities to international organizations. Collaboration enables the sharing of knowledge, technology, and resources to create solutions that benefit all parties involved.

One of the most important examples of global collaboration is the **United Nations Sustainable Development Goals (SDGs)**, which consist of 17 interconnected goals that address various facets of global sustainability, from poverty reduction to clean energy and responsible consumption (United Nations, 2015). These goals foster cooperation between governments, businesses, and civil society to tackle environmental and social challenges together.

International agreements, such as the **Paris Agreement** on climate change, represent another crucial aspect of global collaboration. The Paris Agreement aims to limit global warming to well below 2°C, with countries committing to specific emission reduction targets. The agreement underscores the need for collective action in addressing climate change, with both developed and developing nations playing a role in reducing emissions and adapting to climate impacts (UNFCCC, 2015).

At the corporate level, **corporations** are increasingly partnering with **non-governmental organizations (NGOs)** and **community groups** to address environmental issues. Many companies now work alongside NGOs to ensure that their supply chains are **sustainable** and that they are minimizing their environmental impact, leading to mutual benefits for both the businesses and the communities they serve (World Resources Institute, 2017).

The Path Forward: A Collective Effort for a Sustainable Future

As we look toward the future, it is clear that sustainability is not a singular goal but a multifaceted challenge that requires collective efforts across all sectors of society. From governments and businesses to individuals and grassroots organizations, every action counts in creating a more sustainable world. It is essential that we continue to **innovate**, **educate**, and **collaborate** to address the complex issues facing our planet.

At the heart of these efforts is the recognition that sustainability is intertwined with social justice, economic development, and environmental protection. Achieving a truly sustainable future requires addressing **inequalities**, promoting **inclusive growth**, and ensuring that no one is left behind in the pursuit of a greener and more equitable world.

By continuing to promote **sustainable policies**, **green technologies**, and **conservation efforts**, we can mitigate the environmental challenges that threaten our planet. However, these actions must be underpinned by a **shared vision**, grounded in **collective responsibility**, and supported by **international cooperation**. Only through such collective action can we ensure that future generations inherit a planet capable of sustaining life and providing opportunities for all.

The path to a sustainable future is undoubtedly complex, but with continued innovation, collaboration, and dedication, it is within our reach. It is a challenge that requires us all to take responsibility for our actions, contribute to solutions, and work together toward a better world.

100 recommendations for saving our planet

Here are 100 recommendations for saving our planet using modern technology:

Energy & Renewables

1. Invest in **solar energy** for homes and businesses.
2. Expand **wind farms** for sustainable electricity.
3. Develop **fusion energy** as a clean power source.
4. Improve **battery storage** for renewable energy.
5. Increase **hydropower efficiency** with smart turbines.
6. Use **geothermal energy** for heating and cooling.
7. Implement **wave and tidal energy** systems.
8. Deploy **floating solar farms** on water bodies.
9. Research and scale up **hydrogen fuel technology**.
10. Implement **smart grids** to optimize energy distribution.

Transportation & Mobility

1. Expand **electric vehicle (EV) infrastructure**.
2. Develop **solid-state batteries** for EVs.
3. Promote **hydrogen-powered vehicles**.
4. Expand **public transportation** using clean energy.
5. Use **AI to optimize traffic flow**, reducing emissions.
6. Encourage **micro-mobility solutions** (e-bikes, e-scooters).
7. Develop **hyperloop technology** for long-distance travel.
8. Improve **biofuels for aviation.**
9. Implement **autonomous electric buses** in cities.
10. Support **drone delivery systems** to reduce road congestion.

Agriculture & Food Tech

1. Promote **vertical farming** in urban areas.
2. Use **AI-powered irrigation** for water efficiency.
3. Expand **lab-grown meat** to reduce deforestation.
4. Improve **precision farming** with IoT sensors.
5. Utilize **robotic pollinators** to aid declining bee populations.
6. Develop **plant-based alternatives** for dairy and meat.
7. Use **CRISPR gene editing** to create drought-resistant crops.
8. Reduce **food waste** using blockchain for supply chain tracking.
9. Implement **smart greenhouses** for year-round farming.
10. Use **seaweed farming** to absorb carbon dioxide.

Waste Management & Recycling

1. Invest in **AI-powered waste sorting robots.**
2. Expand **chemical recycling** for plastics.
3. Implement **smart trash bins** that notify collection services.
4. Develop **biodegradable plastics** from plant-based materials.
5. Use **enzymes to break down plastic waste.**
6. Encourage **waste-to-energy plants.**
7. Improve **e-waste recycling** for electronic components.
8. Develop **3D printing with recycled materials.**
9. Use **blockchain for waste tracking.**
10. Promote **zero-waste manufacturing.**

Water Conservation & Purification

1. Deploy **solar-powered water desalination.**
2. Use **AI to detect water leaks** in urban systems.
3. Implement **atmospheric water generators** to collect moisture.
4. Improve **greywater recycling systems.**
5. Expand **nanotechnology for water purification.**
6. Use **drone mapping** for monitoring water pollution.
7. Implement **smart irrigation systems** in agriculture.
8. Increase **ocean cleanup efforts using AI-driven vessels.**
9. Promote **rainwater harvesting technology.**
10. Develop **self-cleaning water filtration membranes.**

Climate Monitoring & Carbon Capture

1. Expand **direct air capture (DAC) technology.**
2. Use **AI-driven climate modeling** for better predictions.
3. Implement **carbon sequestration in soil.**
4. Develop **algae-based CO_2 absorption systems.**
5. Invest in **biochar technology** for soil carbon storage.
6. Use **satellites for real-time deforestation tracking.**
7. Promote **reforestation with drone seeding technology.**
8. Deploy **floating artificial islands for carbon capture.**
9. Improve **weather modification technology** to prevent droughts.
10. Enhance **blue carbon initiatives** (mangroves, seagrass).

Smart Cities & Sustainable Living

1. Build **green smart buildings** with AI efficiency controls.
2. Use **rooftop gardens** for natural insulation.
3. Implement **energy-efficient street lighting.**
4. Develop **self-sufficient eco-communities.**

5. Promote **smart home automation** to reduce energy waste.
6. Expand **urban afforestation projects.**
7. Use **AI-driven disaster response systems.**
8. Integrate **circular economy principles** in urban planning.
9. Install **air-purifying urban structures.**
10. Develop **biophilic architecture** for healthier cities.

Biodiversity & Wildlife Protection

1. Use **AI-powered wildlife monitoring** to track endangered species.
2. Develop **gene banks for preserving genetic diversity.**
3. Implement **automated anti-poaching drones.**
4. Expand **coral reef restoration with 3D printing.**
5. Use **satellite tracking for illegal deforestation.**
6. Improve **marine biodiversity monitoring with underwater drones.**
7. Implement **smart fishing nets** to prevent bycatch.
8. Develop **seed-spreading robots** for reforestation.
9. Use **AI-powered image recognition** to detect illegal wildlife trade.
10. Expand **eco-friendly tourism initiatives.**

Green Manufacturing & Materials

1. Develop **self-healing materials** to reduce waste.
2. Use **3D printing for sustainable construction.**
3. Implement **bio-based alternatives to cement.**
4. Expand **graphene-based energy storage.**
5. Develop **biodegradable electronics.**
6. Promote **self-sufficient factories** powered by renewables.
7. Use **AI for sustainable supply chain management.**

8. Develop **modular, recyclable construction materials.**
9. Implement **zero-emission steel production.**
10. Expand **sustainable textile manufacturing.**

Artificial Intelligence & Big Data for Sustainability

1. Use **AI to optimize energy consumption** in industries.
2. Implement **blockchain for carbon credit trading.**
3. Improve **climate risk assessments** using AI.
4. Deploy **AI-driven urban air quality monitoring.**
5. Use **machine learning for wildlife conservation.**
6. Optimize **global food supply chains with big data.**
7. Use **satellites and AI to predict natural disasters.**
8. Expand **AI-driven deforestation detection.**
9. Implement **AI for smart waste collection.**

100. Develop **personalized carbon footprint tracking apps.**

By integrating these **modern technologies**, we can **protect our planet, reduce waste, and create a sustainable future.**

References

1. Aragão, L. E. O. C., et al. (2018). 21st century drought-related fires counteract the decline of Amazon deforestation carbon emissions. *Nature Communications, 9*(1), 1-12. https://doi.org/10.1038/s41467-018-02990-1
2. Bastin, J. F., et al. (2019). The global tree restoration potential. *Science, 365*(6448), 76-79. https://doi.org/10.1126/science.aax0848
3. Boucher, D., et al. (2018). *Ending tropical deforestation: The global debate about biofuels and land-use change.* Union of Concerned Scientists.
4. Bullard, R. D. (2005). *The quest for environmental justice:*

Human rights and the politics of pollution. Sierra Club Books.

5. Cardinale, B. J., et al. (2012). Biodiversity loss and its impact on humanity. *Nature, 486*(7401), 59-67. https://doi.org/10.1038/nature11148
6. Coady, D., et al. (2019). Global fossil fuel subsidies remain large: An update based on country-level estimates. *IMF Working Paper No. 19/89*. https://doi.org/10.5089/9781513512269.001
7. Cohen, J. E. (2006). Human population: The next half century. In J. E. Cohen, et al. (Eds.), *Science magazine's state of the planet 2006-2007* (pp. 13-21). Island Press.
8. Costanza, R., et al. (1997). The value of the world's ecosystem services and natural capital. *Nature, 387*(6630), 253-260. https://doi.org/10.1038/387253a0
9. Ellis, E. C., et al. (2010). Anthropogenic transformation of the biomes, 1700 to 2000. *Global Ecology and Biogeography, 19*(5), 589-606. https://doi.org/10.1111/j.1466-8238.2010.00573.x
10. EPA. (2020). *Inventory of U.S. greenhouse gas emissions and sinks*. Environmental Protection Agency. https://www.epa.gov/ghgemissions
11. FAO. (2020). *Global forest resources assessment 2020*. Food and Agriculture Organization of the United Nations. https://doi.org/10.4060/ca8753en
12. Fearnside, P. M. (2005). Deforestation in Brazilian Amazonia: History, rates, and consequences. *Conservation Biology, 19*(3), 680-688. https://doi.org/10.1111/j.1523-1739.2005.00320.x
13. Foley, J. A., et al. (2011). Solutions for a cultivated planet. *Nature, 478*(7369), 337-342. https://doi.org/10.1038/nature10452

14. GEF. (2020). *Global Environment Facility annual report 2020*. Global Environment Facility.
15. Gunter, S. M., et al. (2019). Environmental impacts of renewable energy technologies. In *Renewable energy and environmental sustainability* (pp. 45-68). Springer.
16. Hansen, J., et al. (2008). Target atmospheric CO2: Where should humanity aim? *Open Atmospheric Science Journal, 2*, 217-231. https://doi.org/10.2174/1874282300802010217
17. IEA. (2020). *Renewables 2020: Analysis and forecast to 2025*. International Energy Agency. https://www.iea.org/reports/renewables-2020
18. IOM. (2020). *World migration report 2020*. International Organization for Migration.
19. IPCC. (2021). *Climate change 2021: The physical science basis*. Contribution of Working Group I to the Sixth Assessment Report of the Intergovernmental Panel on Climate Change. https://doi.org/10.1017/9781009157896
20. IRENA. (2020). *Global renewables outlook: Energy transformation 2050*. International Renewable Energy Agency. https://www.irena.org/publications/2020/Apr/Global-Renewables-Outlook-2020
21. IUCN. (2020). *The IUCN red list of threatened species*. International Union for Conservation of Nature. https://www.iucnredlist.org
22. Jambeck, J. R., et al. (2015). Plastic waste inputs from land into the ocean. *Science, 347*(6223), 768-771. https://doi.org/10.1126/science.1260352
23. Lal, R. (2004). Soil carbon sequestration impacts on global climate change and food security. *Science, 304*(5677), 1623-1627. https://doi.org/10.1126/science.1097396

24. Lund, J. W., & Boyd, T. L. (2016). Direct utilization of geothermal energy 2015 worldwide review. *Geothermics, 60*, 66-93. https://doi.org/10.1016/j.geothermics.2015.11.001
25. McAdam, J. (2012). *Climate change, forced migration, and international law*. Oxford University Press.
26. NASA. (2020). *Global climate change: Vital signs of the planet*. National Aeronautics and Space Administration. https://climate.nasa.gov/
27. National Geographic. (2020). *Planet or plastic?* National Geographic Society. https://www.nationalgeographic.com/environment/plastic
28. NOAA. (2021). *National Oceanic and Atmospheric Administration state of the climate reports*. https://www.noaa.gov/state-of-climate
29. Pauly, D., et al. (2002). Towards sustainability in world fisheries. *Nature, 418*(6898), 689-695. https://doi.org/10.1038/nature01050
30. Pechony, O., & Shindell, D. T. (2010). Driving forces of global wildfires over the past millennium and the forthcoming century. *Proceedings of the National Academy of Sciences, 107*(45), 19167-19170. https://doi.org/10.1073/pnas.1003669107
31. Pimm, S. L., et al. (2014). The biodiversity of species and their rates of extinction, distribution, and protection. *Science, 344*(6187), 1246752. https://doi.org/10.1126/science.1246752
32. Postel, S. L. (1999). *Pillar of sand: Can the irrigation miracle last?* Worldwatch Institute.
33. Richter, B. D., et al. (2019). Water scarcity and fish imperilment driven by beef production. *Nature*

Sustainability, 2(4), 319-328. https://doi.org/10.1038/s41893-019-0170-0

34. Rockström, J., et al. (2009). A safe operating space for humanity. *Nature, 461*(7263), 472-475. https://doi.org/10.1038/461472a
35. Sachs, J. D. (2015). *The age of sustainable development*. Columbia University Press.
36. Sala, O. E., et al. (2000). Global biodiversity scenarios for the year 2100. *Science, 287*(5459), 1770-1774. https://doi.org/10.1126/science.287.5459.1770
37. Seto, K. C., et al. (2014). Human settlements, infrastructure and spatial planning. In *Climate change 2014: Mitigation of climate change* (pp. 211-320). Contribution of Working Group III to the Fifth Assessment Report of the Intergovernmental Panel on Climate Change.
38. SF Environment. (2020). *Zero waste annual report*. San Francisco Department of the Environment.
39. Steffen, W., et al. (2011). The Anthropocene: From global change to planetary stewardship. *Ambio, 40*(7), 739-761. https://doi.org/10.1007/s13280-011-0185-x
40. Swedish EPA. (2020). *Sweden's climate policy framework*. Swedish Environmental Protection Agency. https://www.naturvardsverket.se/en/
41. Tilman, D., et al. (2011). Global food demand and the sustainable intensification of agriculture. *Proceedings of the National Academy of Sciences, 108*(50), 20260-20264. https://doi.org/10.1073/pnas.1116437108
42. Ueda, K., et al. (2018). Waste management and recycling in Japan. *Journal of Material Cycles and Waste Management, 20*(1), 1-12. https://doi.org/10.1007/s10163-017-0707-x

43. UNEP. (2017). *The Montreal Protocol on substances that deplete the ozone layer*. United Nations Environment Programme. https://ozone.unep.org/
44. UNEP. (2019). *Global environment outlook – GEO-6: Healthy planet, healthy people*. United Nations Environment Programme. https://www.unep.org/resources/global-environment-outlook-6
45. UNFCCC. (2015). *Paris Agreement*. United Nations Framework Convention on Climate Change. https://unfccc.int/process-and-meetings/the-paris-agreement/the-paris-agreement
46. WHO. (2021). *WHO global air quality guidelines: Particulate matter (PM2.5 and PM10), ozone, nitrogen dioxide, sulfur dioxide and carbon monoxide*. World Health Organization. https://www.who.int/publications/i/item/9789240062067
47. World Bank. (2018). *What a waste 2.0: A global snapshot of solid waste management to 2050*. World Bank Group. https://www.worldbank.org/en/publication/what-a-waste
48. WWF. (2020). *Living planet report 2020 - Bending the curve of biodiversity loss*. World Wildlife Fund. https://www.worldwildlife.org/publications/living-planet-report-2020
49. Zhao, M., et al. (2019). The role of deforestation in climate change. *Environmental Research Letters, 14*(12), 123005. https://doi.org/10.1088/1748-9326/ab5659
50. Park, J., et al. (2020). Waste management in South Korea: Policies and challenges. *Waste Management & Research, 38*(8), 787-795. https://doi.org/10.1177/0734242X20923515

Acknowledgment

"This book exists because of the extraordinary people who believed in it—and in me. To my [family, whose unwavering support was my foundation; to my friends/colleagues, who kept me sane); and to every unseen hand that guided this journey—thank you. None of these words would matter without you."

Jihad Saffouri

2025

Don't miss out!

Visit the website below and you can sign up to receive emails whenever Jihad Saffouri publishes a new book. There's no charge and no obligation.

https://books2read.com/r/B-A-AJQRD-BINFG

BOOKS 2 READ

Connecting independent readers to independent writers.

About the Author

Jihad Mohammad Ibrahim Saffouri is an educator, environmental advocate, and visionary thinker . He has a a deep-rooted commitment to sustainability, human development, and education, Jihad has spent decades inspiring students, colleagues, and readers through his work. His multidisciplinary expertise spans environmental science, psychology, and educational reform, making his voice both timely and essential in today's climate-conscious world.

Driven by a passion for justice and a belief in the power of knowledge to create change, he wrote *"Saving Our Planet: A Call to Action for a Sustainable Future"* as both a wake-up call and a hopeful guide. His insights blend scientific understanding with ethical urgency, aiming to empower individuals and communities to take meaningful steps toward environmental preservation and a more sustainable world for future generations.

www.ingramcontent.com/pod-product-compliance
Lightning Source LLC
LaVergne TN
LVHW090304210926
848480LV00001B/129